top cop joins with a former **Green Beret** to teach you

AMMO FOREVER

The Complete What To Shoot & How Manual For Rifles and Shotguns

<u>1st Editon</u>

Book I

by

Don Paul
&
David Blaisdell Smith

About the authors: David Blaisdell Smith is a California Highway Patrol weapons specialist and marksmanship instructor. He has written numerous magazine articles on guns and hunting for all kinds of magazines. Don Paul is the former Green Beret who became a writer after his parachute failed over Panamanian jungles in 1976. He's the author of several outdoor survival books.

Cover: Alan Iglesias of Escondido, a leading 3-D artist in the United States. Concept of ammunition in a bubble without guns and beclouded depicts how things were. Now, get the complete story, ammo, guns and shooter, all improved to outshoot all others.

Gratitude: To the manufacturers of guns, holsters and reloading equipment, who's products help us to remain free and independent. Owning some of what they produce is one giant step toward being your own free person.

A.M.D.G.

Prayer: Before we publish, Dave and I bow our heads to acknowledge God, thank Him for his blessings, and ask not that He helps us write, but that we conform to His Word and Will as we do. *Lord, we rededicate everything we are to you as this book goes to publication. Our lives are yours, our works are yours, and we acknowledge your Holy Scripture as standing forever. We believe our world is under attack by the powers and principalities and we offer this knowledge to fellow citizens of a country you founded. We thank you for what our beloved country once was---and can be---if your people who are called by your name will repent and turn to prayer. Amen.*

Library of Congress Catalog Card Number: 94-69563

Publisher's Cataloging in Publication

Paul, Don, 1937-

Ammo forever : the complete what to shoot and how manual for rifles and shotguns / by Don Paul and David Blaisdell Smith. -- 1st ed.

p. cm.

Includes index.

Preassigned LCCN: 94-69563.

ISBN 0-938263-15-3

1. Firearms. 2. Shotguns. 3. Rifles. 4. Ammunition. 5. Survival Skills. 6. Shooting. I. Title.

TS534.5.P38 1995 623.4'42

QBI94-21211

Introducing. . .

QUICK-READER BOOKS by PATH FINDER

New-method, easy-to-read, how-to books.

To write any good how-to book, you have to know your topic thoroughly. That takes years of experience, long hours of reading and special schools. I did all that. I was a Green Beret, a chosen member of Sixth Army's Marksmanship Detachment, a winner of many rifle shooting matches and an avid student. Still many people know more than I. David B. Smith is the best. He probably could have become a high ranking officer in the highway patrol, but he doesn't like desks; he likes guns. So he taught other cops how to shoot. He's a riot gun pro. Often after 15 minutes, his students were blasting right on target. He reloaded all kinds of ammunition and worked as a gunsmith. We at Path Finder consider ourselves blessed to have found him; he's great as a co-author. I could ask him anything, anytime and he knew the answer immediately. In his original manuscript, he gave away many secrets he learned from his dad and other cops---all avid shooters and hunters.

These books are different. Complete, as used in the title doesn't mean we included every little detail. Complete means we tied the perfection of the reload---to the precision of the gun---to the profficiency of the shooter. This book improves all three. Others have published information on reloading, bullet energy, velocities and bench rest accuracy. That's all complicated stuff; we don't do that. But you need to have the whole story---not just piece-meal information. By the time you finish reading this, you and your firearm will be able to out-shoot anything else out there. That's because we take you from cartridge, super accurate and handmade, to firearm, specially treated and accurized. Then we show you how to shoot it like a pro---from a longer distance with super precision.

At Path Finder, we have a different publishing philosophy. We take our best shot at going beyond any other knowledge in a given field. In this book we let out secrets of new shooting developments. We also dispel some of the myths about firearms and shooting. In swift, easy-to-understand detail, you'll discover the newest methods for placing a bullet with precision. Note: Some dangerous information not included, but will be made available to military and police on request.

> A technical book on bullets, guns and shooting such as this must be computer published, otherwise it takes too long to get an update to our readers.

Though this information is critical, the text is written and specially formatted for speed reading. We designed this text so you get all the information---fast. We really care about your time. This book contains exactly 44,068 words. Hopefully, we didn't waste even one. Our electronic scrubbers report:

If you sub-vocalize when you read at 550 wpm, you should complete this whole book in 1 hour and 19 minutes. After electronic scrubbing, we achieved the reading-ease of Hemingway's short stories. We average under 1.5 syllables per word. Average sentence: Under 15 wds.

Path Finder began over 14 years ago. We first invented a way to keep you from getting lost in the woods without using a map; it was called, *Never Get lost. The Green Beret's Compass Course.* Over 40,000 copies are in print. After that, we added to our book list and widened our distribution. We published:

Everybody's Outdoor Survival Guide

Great Livin' in Grubby Times best selling survival book.

Everybody's Knife Bible over 30,000 copies!

24 + Ways to Use Your Hammock in the Field.

SECURE FROM CRIME, How To Be Your Own Bodyguard.

We develop and write about new ideas. We're the innovative people who wrote about outdoor know-how and discovered:

√A two-ounce, 30¢, wilderness bed for sleeping above ground.

√The light for your hunting knife sheath so the ground is illuminated for you in the jungle or woods at night.

√Paint for the bottom of your outdoor boots to make them slippery. Wear the paint off the bottom; they no longer clog.

√How to use animals to double your survive-ability.

√Green Beret team concepts applied to survival groups so you can enjoy the ultimate life-style outdoors.

√Never become a victim. New and clever ways to beat criminals. ***SECURE FROM CRIME*** is so popular, the authors have appeared on over 200 media shows.

All our books have gone into multiple editions. Most major outdoor magazines have reviewed our books and our systems have been adopted by many outdoor organizations. Once you own any of our books, you can order a new, updated copy for half price.

(See the order coupon in the back of this book.)

Without this book you may run out of good ammo, the guns to shoot it, or the knowledge to put it all together. Dave Smith brings over 30 years experience into this subject as a firearms expert and instructor for the California Highway Patrol. Don Paul is a former Green Beret who has published eight how-to books. They both reveal a lot of new gun and ammunition data in this book.

AMMO FOREVER

THE COMPLETE
WHAT TO SHOOT AND HOW MANUAL

by

Don Paul & David B Smith

Book I

CONTENTS

Section III
MODERN RIFLE

Section IV
WAY TO PRACTICE—-SHOOTING HOT AIR

APPENDICES

AMMO FOREVER...

The Complete What To Shoot and How Manual

by

Sgt. Don Paul and David B. Smith

Section I Chapter 1

WHY THESE BOOKS?

Path Finder won't publish anything if it isn't new and unique. We wrote these because we saw a big empty gap in the gun, ballistics and bullet placement information world. Other books on shooting by Green Berets are fine, notably John Plaster *(ULTIMATE SNIPER)* and Chuck Karwan *(COMBAT HANDGUNNING).* These guys know a lot; the only thing I have against them is they were Army officers in Special Forces. As anybody knows, enlisted people make everything happen in the military. I also like Dean Grennell on reloading. He's the boss. He knows all the wildcats, CUP pressures, headspacing dimensions and bullet velocities---probably by heart.

Most reloading books say you can either reload (for quantity) or hand load (for quality). We think you can do both. The approach we take is to make super ammo. Then improve your weapon to shoot super ammo as it was intended to be shot. Finally, use some of the tricks we know to become a super shooter. In one statement: These books teach you to develop magic rounds for your much improved individual weapon which you will shoot like a super marksman.

> It annoys us that nobody has ever taught what we consider to be so obvious before.

Contrary to popular belief, new weapons---"new-in -box"--- are not in perfect shooting order and you can (and should) improve them. Why go out into the field with a firearm that only shoots as well as the rest of them? Why is this the case with new guns? It's because the manufacturers in this country might be brave, but they cower in fear when they hear "lawsuit". I can't blame them. I wouldn't improve my product one bit if it opened me up to product liability. Another reason to be careful with innovation is the anti-gun lobby. Even though most pistols are used at night, manufacturers must make a "sporting" weapon for hunting in the field. Most magazines take the same position. Finally your guns don't come with certain improvements because labor costs so much. You may find twenty hours to work on your weapon. Even if those minutes are precious, you never call in sick, go on strike, or demand perks. If you don't mind working for yourself, "new-in-box" means you'll find lots of room for making your weapon first class.

After you do that, learn to shoot well. Everybody knows the basics, but the use for which each weapon was designed plus the way it operates and the distance it covers from the muzzle—-all—-require different shooting techniques. That's something we really get into in this book. We came up with a wobble detector, which tells your shooting variation in MOA. Once you have that information, you know at precisely what distance you can hit. Even more important, you can copy the page on which we drew the graph and work on your shooting positions **while you study your wobble.** Never before has that been done. In the past all we had to go by were the bullet holes you made in the target. But the essence of good marksmanship (or any other sport demanding physical performance) is in tracing error to cause. Only the detector enables you to reduce wobble by correcting your position until you can hold tightly enough to shoot super groups.

Result: A more accurate shooter with superior gun and better ammo can out-shoot the world—-farther and more precisely with bullets that have much more influence on any target. Other books don't tie the three subjects—-reloading, improvement and shooting—-together. Nevertheless, if you want to out shoot the rest of the world, you **must** do that. To write the "Complete What to Shoot and How Books," you have to say this: **"Buy this kind of weapon, improve it this way, put together some special ammunition and make it out shoot every other gun out there."**

What good is any bullet if it doesn't fill a need for your own rifle? I never read any reloading material about putting a dime in a 12 gauge load. Some people think of bullets as delivery vehicles for foreign substances. What about making a handgun bullet create more than one kind of wound in the target body? You can't make that happen unless you adjust the powder load to give you the required velocity at the distance to target you will encounter. That's something store-bought ammo will never do!

Why? Because some attorney will sue for millions if a manufacturer doesn't stay within super-safe limits.

Furthermore, this makes us really mad: Most of the gun magazines and books ignore the problem of incoming. People who want life saving knowledge (such as cops and military forces) read gun literature and go out to war (either against a foreign enemy or a local criminal enemy) and die because they didn't get the whole story. Why has this never been done? Because it isn't fun reading about getting holey. Getting shot at is something nobody enjoys thinking about. However, if you never consider the possibility, you'll never be prepared. I guarantee you, parts of these books won't be fun to read but then you won't go out there half cocked, either. If you would like to talk to most of the people who went into combat only while shooting and didn't consider incoming, visit your police cemetery.

These books (Book II covers Handguns) will take you from the reloading bench, into the chamber, down the barrel, and into the target with amazing results. If you own a shotgun now, you can buy ammo that will do a nice job of shooting birds and small game at close range. But we're going to stretch this weapon out for you so it performs a variety of unusual tasks. Your rifle may already have brought some meat home from 300 yards away, but when you add a few improvements to the weapon and load it with some fancy stuff, "king of the woods" will be an understatement for the respect you'll command. Your handgun will take care of defense nicely with store-bought ammo, but when you come away from our reloading bench with a fire-lapped barrel, it will be the devastator for all seasons. (Book II)

These two volumes are not books about the complications of handloading. We don't explain intricate things like *CUP*— cupric units of pressure, which is an old way of measuring pressure build-up in your weapon's chamber. We also don't list loads. Someone else already did that.

> We recommend you read DBI on shotshell reloading. They list some of the best loads in the back of the book, and they have pictures of every possible product you can buy.

When I read others, I find most of them approach the subject of reloading without defining targets. In addition to the targets you find afield, we consider targets to be the kinds of animals who prowl around in our cities' concrete jungles. As a matter of fact, we consider how dangerous it can be to practice shooting in the woods or desert near a large city, where other gangs may be practicing. Remember, Path Finder published ***SECURE FROM CRIME, How To Be Your Own Bodyguard,*** and in the research for that book I learned you are never safe anywhere. Be heads up. Guns are now precious commodities. Gunstores have sold more than ever before. After Clinton, there was a national shortage of primers. Many gangbangers would kill for guns. Be careful out there.

We want your guns to speak with **ultimate authority!** You can't make that happen unless you reload to achieve a precise result. Great handloading means little if you don't know how it relates to weapon performance. Improved and super accurate guns mean little if you can't shoot them well. What other books **don't do** is tie the reload to weapons' performance. To fix that, we designed an all-new form for both shotguns and rifles. First you record reload data and code the ammunition to the page. Later, you record how the load performed on the same page. Starting with the hole in the target, we will take you back to the round in the chamber. That's why you'll learn shooting techniques, weapon performance capabilities and tactical use of your guns, in addition to reloading. Why put together a flechette round when you don't know what it will do for you? What compound might someone install in a hollow point bullet?

Path Finder has been after me to produce these books for a long time. I already have eight books published, and I am busy promoting the most recent one called ***SECURE FROM CRIME.*** To date, I've appeared on over 200 radio and TV shows; I think I can answer any question about how to avoid becoming a victim.

I think I could have written some of what's in these two books on my own. It might have been *good.* But with Dave Smith, a retired highway patrolman and gun expert with a lot of experience in making all kinds of cartridges, I could produce a *great* book. I thought I knew a lot. Now I suspect he forgot more than I'll ever know. The two of us writing together have probably produced the best books Path Finder has released to date.

I Chapter 2

BASIC GUN IMPROVEMENT

Later, we deal with improving different kinds of guns. But for now, almost all new guns have some common deficiencies you wouldn't discover. What gripes us is this: Nobody tells you. The manufacturers' advertising people certainly don't want to talk about problems (something wrong?) and the magazines and publications focus on the good stuff (otherwise they wouldn't get advertising money). Guns don't need huge manufacturing mistakes to be inferior either. Even when your new gun is the best the factory turns out, it needs improvement.

New guns aren't up to par. Anything extra a factory worker does to improve your gun can cost astronomically. So to get the ideal finished product, you have to do it yourself. Lovingly. Spend a little time with this gun and it will provide you with super service for the rest of your life.

THE FIRST THING---CHECK IT OUT.

While working as a gunsmith, Dave has heard some unbelievable complaints: "This barrel is crooked." "This pistol has a smooth bore." Surprise. The complaints were valid and the gun went back to the factory. What's the rule?

When you buy a new gun, check it. Most guns will be OK, and those which aren't go right back to the factory. Just make sure yours is in the "OK" group. Look at the stock. Dented? That means rough handling during shipment. Check shotguns for dents in either the magazine or barrel. Use the bore light at the store and look for bends in the barrel. Twist the choke down and watch it constrict. Look OK? Has the rib been soldered on properly. Jiggle it. Snap the trigger. We don't suspect anything will be wrong with the firing pin, but if you place a dime on the bolt face with the muzzle pointing skyward and snap the trigger, the dime should jump inside. Hear that? For both rifles and shotguns, try to snap the trigger with the safety on. Then put the safety on with the weapon cocked and tap the butt on the ground a few times. Will this weapon fire with the safety on when dropped? For rifles, inspect the bore carefully with a strong bore light. Use a powerful magnifying glass to look at the crown. I want each land in the crown to be perfect---not deformed or chipped. Take out a dollar bill and slide it under the barrel at the front of the forearm. Then work it all the way back to the action. If the barrel isn't free floated, the stock needs some work.

BREAKING IN A NEW BARREL

NIB—-New in Box is alluring to every gun buyer. As with a new car, however, break it in gently. Let's look at what a bullet does in a new barrel as the powder pushes it, faster and faster, down the twists inside and out the muzzle. More dedicated shooters will at least run a patch or two through the bore. The first patch will be soaked in cleaner. The second might have lightweight oil on it. When the patch is clean, the gun is good to go. Some shooters merely pop in a new cartridge, hold the weapon tightly, and "let the first one fly." That's bad news.

HOW TO RUIN A BARREL---A THEORY

As the bullet gains speed, the lands cut into the bullet jacket and cause the bullet to spin as it slides along the grooves. Let's suppose a microscopic piece of metal (steel) lies in one of those grooves. At a speed around 1,000 miles per hour, that tiny chip gets picked up by the nose of the bullet. If the bullet nose were flat, the chip might simply fly forward. But that's not so. The bullet edges are round, so the chip gets pushed down hard into the barrel's grooving and gouges a slight canal as it flies out ahead.

In a new barrel chances are fair that other chips may dislodge and lie in wait for other bullets to do the same. Don't worry; we have a cure for the problem. This is what you do. Clean the bore; then shoot; clean again; then shoot—-about ten times for any new barrel. After that, keep your barrel clean. The new bore cleaners are the best you can buy; use them. Take a close look at the patches after you pull them out of the barrel. If the patch isn't clean, then the barrel is still dirty.

Suppose you don't clean yours. Then all the chips laying in wait to be pressed into the side of your barrel with tremendous force will gouge the grooves. Those new little gouges allow tiny amounts of hot gas to escape around the sides of the bullet every time you shoot. The effect on the bullet is about the same as a poor crown job. Like a golfer getting "gotchas" just as he putts, your bullet gets a little hot-gas shove sideways just as it leaves the muzzle.

Straight barrel with bumps

POLISHING THE ROUGH SURFACES AND REMOVING THE HIGH SPOTS

FIRE LAPPING

I don't believe I would own a new gun unless I paid the extra money for the NECO kit and fire-lapped the barrel. To shoot best the barrel must have no narrow spots or rough surfaces inside. Fire-lapping is just about the only method you can use to accomplish all this. Once the bore is polished smooth, you have just made sure your gun will hit at distances far better than other, ordinary guns.

Whether a handgun or rifle, fire lapping is critical if you want to shoot accurately. However, if you want accuracy now and value later, fire lapping is both critical and necessary.

FREE FLOAT AND BEDDING

Let us take you to a fictional place inside your barrel's steel. As the bullet twists out of the barrel, the barrel twists in the opposite direction. (Newton's Law). Some shooters call it barrel "whump" and they eliminate most of it by using a bull barrel with enough metal on it to control the whump.

Bull barrels weigh too much for most sporting applications, so manufacturers produce thinner barrels and merely allow the whump to take place. That's OK. But it is not OK if the whump is different every time you shoot.

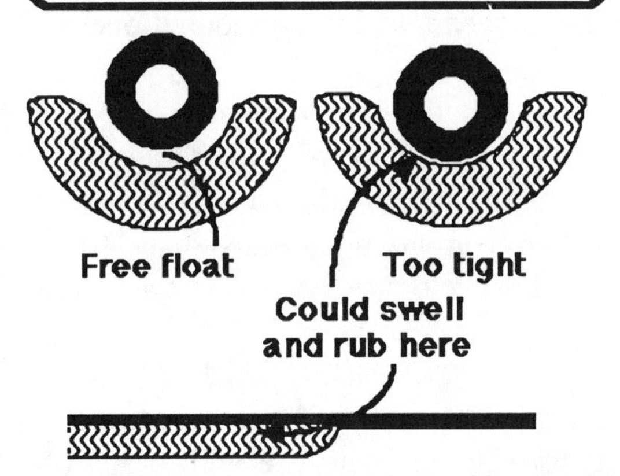

What causes the whump to change? A rifle stock rubbing up against the barrel. Wood stocks have a nasty habit of doing that, especially when they absorb moisture from the atmosphere. Therefore, make sure your barrel doesn't touch your stock. Float the barrel up and away from the stock enough so you can slide a dollar bill all the way back to the chamber. Check it. You may have to remove some more of the stock's wood. Also, as soon as you can, seal the stock with Linspeed Oil so moisture can't get into the wood and cause it to swell. Several other stock sealers are sold in gunstores everywhere.

If the action slides back into the stock when you shoot, then it will move in a variety of directions when it recoils to make your bullet visit some strange places. Marry the action to the stock by glass bedding the stock with strong support. When you do this, of course, free float the barrel.

EXTEND YOUR BARREL'S LIFE

All barrels wear out from firing and have a practical accuracy life. That life depends on what kind of cartridges you fire, the choice of steel for the barrel and the method of manufacture. In order to get some real facts on barrel life, I contacted "Boots" Obermeyer of Obermeyer Rifle Barrels. He has manufactured top grade rifle barrels for years and is considered one of the best in the country.

To wear out the lands in your barrel, shoot high velocity bullets with a lot of bullet drag—-long sides of the bullet rubbing against the barrel. On the other hand, keep your barrel in sharp shooting shape for a long time by shooting low velocity (less powder) bullets with less bullet drag.

BARREL QUALITY

Attorneys have proved that better steel in barrels is dangerous for shooters who fire with barrel obstructions. To reduce products' liability, manufacturers need barrels that will bulge rather than blow apart (flying steel fragments) in the event of accidents. Therefore, most factory made hunting rifles have button-rifled barrels. The rifling is shallow—-the steel soft.

Boots' good target barrels start loosing accuracy between 1500 and 6000 rounds, depending on what cartridge and powder load he shot through them. He went to Chrome-moly steel for awhile, because of its resistance to abrasion. Burned chamber throats still occurred, however, which produced high vertical shots. (A little hot-gas push on the bottom.) He then tried stainless steel. The reward: Three to four times the barrel life. If you shoot a lot, then barrel life is important. On a decent hunting rifle, some shooters put out fewer than twenty rounds a year, which is 800 rounds in forty years. Most barrels will make it.

MAGNUM EXPECTATIONS

Magnum rifles are wonderful because they reach out farther than others, but they lose significant accuracy after 5,000 magnum rounds. Therefore, hot loads shot continually are not a good idea for the survivalist. What's the answer? Load down. Practice and shoot with cartridges using less powder and lighter bullets. When you need the occasional long reach, pull out your heavy duty blasters. Your barrel will thank us.

TOWARD YOUR WEAPONS' LONG LIVES

Buying a new gun? We sit here at this computer and lift a glass to toast: "May you and your weapon live long!" But you will have a lot more to do with it than we will. The key to any firearm's accuracy and long life is the way you take care of it. Everything you use needs a little T.L.C. When you buy your gun, get a soft metal ram rod, (one you can break down and carry with you) liquid cleaner and lubricant, and a set of good tools. Screw drivers which don't fit exactly into screw slots leave little gouges on the screw head, which tell any experienced eye that some amateur worked on your weapon.

Bore cleaners are hi-tech now. Be careful, for some will eat the bluing right off your gun. A friend of mine bought a used target rifle. It shot OK but the barrel looked dirty, so he started cleaning it with some powerful bore cleaner. He cleaned and cleaned until the bore was bright. When he took it shooting again he found that it had lost its accuracy. Some people are really touchy about cleaning or not cleaning weapons. Some of Don's Green Beret buddies named him after the way he cleans

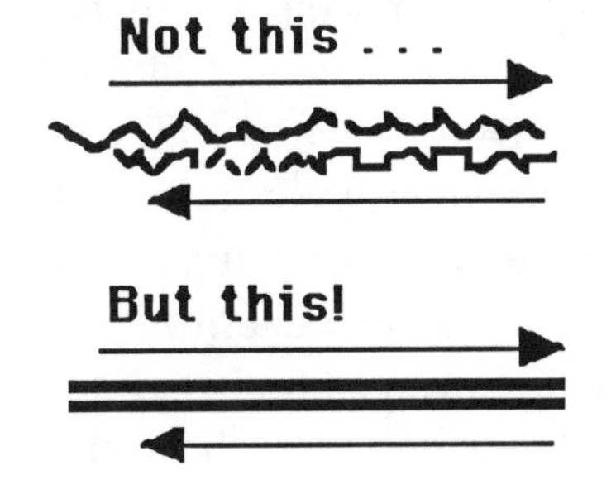

weapons: "PigPen." But one thing Don always did—-and you should also with any automatic weapon operated by gas port pressure—-leave some *BREAK FREE,* or solvent soaking on the gas port overnight. In the morning shoot a round or two to clean the port.

WHERE'S THE RUB, BUB?

Perhaps single shot weapons are the exception, but all others have metal parts which rub and grind *against* one another. After you work your magic, however, they will rub *together.*

Picture two metal surfaces in contact. When they come out of the factory, they are rough; perhaps they only make partial surface contact. They scrape against one another and make harsh noises. But after you finish, they will slide gracefully into position, locking together smoothly in preparation for firing.

How do you make this magic? How will you create smoothness out of roughness, grace out of clumsiness, and silence where there used to be screeching? With a marking pen. On every surface of your new firearm which rubs against another, color it. Then grind it together. Find out where the rub is. If your eyes are only good for long-distance shooting and you need reading glasses, feel free to use a magnifying glass. You'll see right away where the marking has been rubbed off.

Now—-on those surfaces, apply a light rubbing compound. Work the action to let those grinding surfaces wear down until they fit like a hand in a deerskin glove. Careful! Don't overdo it. Use the marking pen over again to tell you when the rub is even. Then stop. Small tiny bits of rubbing compound will stay in the pores of the metal, even after cleaning, so continued use will make your action even smoother.

Once you have the surfaces sliding together, apply

Pro-Tec oil. You won't believe how smooth you've made this new shooting machine. It will make less noise in the field---perhaps a small advantage. But speed is important. Carlos Hathcock could work his bolt as fast as an M-14. I hope you also acquire that kind of speed.

UNCLE DAVE'S FIELD GUNSMITH KIT

Buy those spare parts which have a history of breaking. On a .45 auto pistol, for example, the firing pin should last for life. However, on the new autos the pins are notched for various safety functions and therefore break more easily. Carry a spare. Spare screws for scope mounts and other small applications are likewise a good idea. Shotgunners should carry a spare firing pin. If your choke requires a ring change at the end of the barrel, make sure to bring those rings with you as well as the choke wrench.

An extra extractor is also a good idea for the handloader. Because some high-pressure, handloaded cases get stuck and many shooters try to solve the problem with muscle, you need a spare extractor. No rifle shooter should go into the field without a break down cleaning rod. Without one, the slightest foreign matter can obstruct your barrel. One problem occurred when the hunter dropped his rifle, muzzle down into mud.

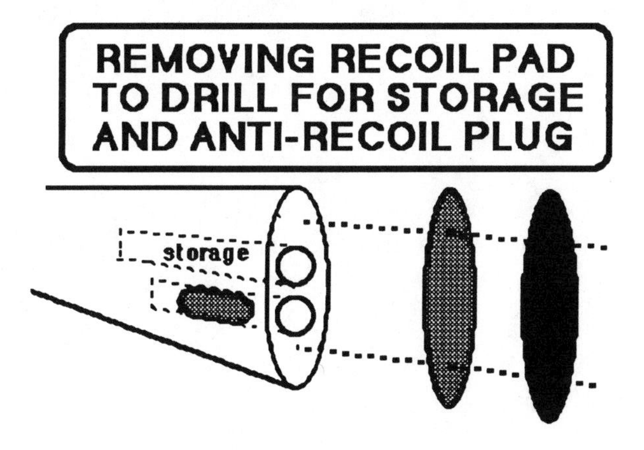

Some tools and cleaning equipment should always go where the weapon goes. A toothbrush is a good idea for reaching into some small crevices. It can also help you coat a bit of oil in hard-to-get places. Will you be adjusting open iron sights in the field? Take a small hammer and brass drift punch.

Store the parts with the weapon. For long guns with wooden stocks, create a storage bin. Remove the butt plate, drill two holes in the stock up toward the trigger. Spare parts go into a plastic wrapper (with Rig or ProTec all over them). Then wrap the plastic in absorbing cloth or cotton from pill bottles and place them snugly in the hole.

N.I.B. or, new-in-box can spell disaster if you simply go out and shoot away. On the other hand, new in the box can mean the best gun you'll ever own, because you didn't inherit somebody else's bullet spraying problem after he neglected to break in the barrel carefully or fire-lap it. Perhaps he even cleaned it with a steel cleaning rod he allowed to rub against the lands at the muzzle. Ouch! With only a little care in the beginning, you and your weapon can do great things for a lifetime.

II-Chapter 1

ABOUT SHOTGUNS

We begin with shotguns for many reasons. The shells are easy to reload. The guns are cheap. Learning to shoot one is easy. They propel a wide variety of projectiles with amazing energy. They make the best home defense weapons (see ***SECURE FROM CRIME***). They also have several other great qualities I listed in the chart in ***GREAT LIVIN' IN GRUBBY TIMES.*** You can modify shotguns with special stocks so they perform in tight quarters (pistol grip) and the variety of shotshell loads handles lots of different challenges.

What's better; a shotgun or a pistol? The key in the whole controversy is this: As pistol range increases, accuracy decreases. With a shotgun, it's the other way around. The farther away you are, the better your chances of hitting---within limits. You can improve your chances by using more powder, more shot and a full choke, but most shotgunners agree that the improvement isn't worth the strain you'll put on the gun---or your shoulder. What pistols do for shotguns is provide great backup. Shotguns are slower to load so they sometimes come up clicking, perhaps misfire, and could fail to feed if you dent the tubular magazine. Therefore, carry a backup handgun. Mine is in a fanny pack with magazines and ammo.

Consider the casing, which we will call the "hull." The modern ones have a brass bottom with a recess for a large primer. Powder (preferably fast burning) is pressured up against the primer flash hole with about 40 pounds of pressure by a wad. That wad consists of a pressure plunger, some shock absorbent plastic rods on top of that, and a shot cup. Inside the cup we drop in the projectiles, either small shot, larger 1/3-inch balls or a slug. Then we crimp the top.

Even though most versatile, the shotgun has limits. The hottest powder load behind a slug can't produce much more range out of a shotgun than a football field and a half. Shotguns are excellent close-range defense weapons because they beat other weapons by a bunch--but only at close (30 yards) range. Shot loads mean you can only be effective at short range—-less than 40 yards with buckshot. People who live in dense woods or jungle don't care about short range, however, because they never get much of a long shot. No matter how you load, you could be in trouble if up against a rifle, or a few handguns which shoot magnum cartridges. A decent rifle person can shoot over 300 yards easily.

Any shot size will kill at close range, but when you move back to 30 or 40 yards, pattern density opens up and smaller shot slows down. This is why: A vacuum forms behind the trailing edge. Only the weight of the projectile keeps it moving. That's why bigger shot (such as #4's) have better terminal whomp than small shot (#8's). If you want more power on target, go to larger shot, which is designated by a smaller number.

Called the fowling piece (for fowl), the scattergun was a favorite weapon in early America. The colonist could load his piece with small shot for birds or use buckshot for deer. Unlike the colonist's fowling piece, we produce today's modern shotguns in several gauges and action types. Also unlike the early pieces, we choke our barrels so we can control the spread of shot. Much like the nozzle on the end of a garden hose, we either shoot a long stream of lead, or we open the choke to improved cylinder so the shot spreads out and covers a greater area (but doesn't travel as far).

Although you can buy shot cartridges for the .22 rimfire, 9mm. rimfire, and pistol cartridges like the .38/.357 revolver, we identify real shotgun shells by gauge—-not caliber. Shotguns come in 410, 20, 28, 16, **12** and 10 gauge. Most popular is the twelve, which is in operation more than any other due to its power and effectiveness.

Shot pellet sizes run from #12 (2,385 to a pound) up to size #2 (90 to a pound). The larger the number---the smaller the shot size. The hulls are mostly plastic and brass-rimmed. Standard case lengths are 2 3/4" and magnums are 3" long. Average muzzle velocity is 1,200 fps. (feet per second).

With buckshot, the higher the number, the larger the shot. (The opposite of birdshot size and numbers). You can choose from BB (.18"), #4 Buck (.24"), #3 Buck

(.25"), #1 Buck (.30"), #0 Buck (.32"), #00 Buck (.33"), and the largest, #000 Buck (.36"). Heavier projectiles use less propellant, which means less velocity. However the heavier the shot load, the more dense the pattern will be. Also, the larger the shot, the better velocity it retains after it leaves the barrel, so it swats your target with more stopping power. We state it this way: Larger shot has better terminal ballistics, which refers to the projectiles' performance **in** the target.

Standard 2 3/4" 12 gauge #00 buck cartridges contain nine .33 caliber lead balls; magnums contain twelve. You can shoot anything out of a shotgun to get some unique results. When we talk about flechette, we refer to the round you can put together yourself by using cut pieces of large paper clips. They will stop an intruder, but they won't fly through a wall and perhaps harm others (children) sleeping elsewhere in the home. Besides birdshot, buckshot and flechette, single projectiles (either copper or lead) are available for the man who makes his own ammo.

RANGE

Whether you use your shotgun to obtain food or go into combat, you have to know at what range (distance-to-target) you can still be effective. We define "range" as how far away a given target is. "Effective range" has long been thought to be a part of the weapon itself. For example, military instructors will tell you the "effective range" of this weapon is so many yards or meters. Well—-that's bull. We wrote the book partially so the military can learn this fact: Range is not an attribute of a weapon alone. It is twofold. A. You determine range by loading your ammunition properly. B. You either increase or decrease range by the way you shoot.

First, use more or better powder when you load so your ejecta reaches out farther. Second, change and improve the projectiles so they slice through the air more effectively. Third, tighten up the nut behind the buttplate (is that me?); practice more with our cheap, junk yard ammo until you really get good at shooting. (Primarily a pistol exercise, see Book Two) Finally, test your ability; then don't try to exceed your ability (the range or distance at which you can consistently hit). That's especially true in combat. When the other guy has long distance shooting capability and you fire farther than you can hit at long range, all you do is disclose your position. Make that mistake and you will have to memorize our system for locating snipers by analyzing incoming fire. (See Rifle Incoming) We'll enable you to locate him so you can send return mail.

Test the ability of various rounds you create. Sometimes you pour in a hot load of powder, increase the weight of the shot and you don't gain much. Generally though, when you increase the size of the shot (by using a lower number) you get more range. Just make sure you get effective range. With shotguns, you find out by using a pattern board.

SLUGGING IT OUT AT A DISTANCE

When rifled slugs became available, accurate range for the shotgun extended beyond 40 yards. With rifling on the outside (vanes) cut into the slug, it spins so it doesn't tumble. Therefore, it lands on target accurately. How far? About goal post---to goal post on a football field. Wheel weights would be about the right consistency if you cast your own slugs. You want them soft in case you forget to open your choke when firing.

Slugs are mostly lead. However, copper-plated lead and solid-copper sabot slugs are now available. Remington's slug is encased in a one-piece sabot, with eight longitudinal slots that open uniformly so the sabot plastic casing separates smoothly from the slug.

Remington claims 2 1/2" groups at 100 yards with this sabot. Rifled shotgun barrels help slug accuracy, but are devastating to any shot patterns, especially buckshot.

Nobody sells sabots for shotgun loaders. But you can buy commercial rounds to keep in inventory. For mix 'n match loaders this should be the last to shoot. The lead part of a sabot is long and bottle-shaped and weighs about an ounce, which enables it to come out of the muzzle at 1500 + fps. Winchester has a new high velocity one ounce sabot slug that moves along smartly at 1,760 fps. I salute.

Here's a safety note: Don't shoot two shotgun gauges at the same time. Mistakes made while doing so frequently cause severe injury. If you don't pay close attention and load a 20 gauge cartridge into your 12 gauge it will slip about halfway down the barrel and lodge. Of course, your next round will push the 20 gauge down the barrel ahead of it—-a little farther. Surprise! The shotgun blows up.

SHOTGUN BARREL
Preparing to explode!

CHOICE OF TYPE IN SHOTGUNS

Shotgun types are as varied as other weapons, but choosing from only a few will keep things simple. You either get two barrels in over/under or side/side, or you get one barrel in a pump or automatic. Sport shooters often go for the two barrels. I don't. I either get a pump or auto, and that's what I want you to get also, but the choice depends on several factors.

AUTOMATIC SHOTGUNS

Autos require maintenance. No grease; no oil; no function. But you can muscle a pump and make it feed and fire. To reproduce ammo for the auto, you have to full-length resize. Otherwise, the gun may not chamber the shells. Another tricky thing with autos: Be careful about the powder you choose. Some powders (flake types) don't burn completely and leave residue in the gun. Flake residue can clog your automatic's gas ports and restrict the amount of gas power available to cycle the bolt. Want to hear about another problem? Autos cycle because of chamber pressure; not load velocity. So you can construct a shell that will produce the same velocity with a different kind of powder, but it won't produce enough chamber pressure to cycle the gun. You may want to reduce your powder charge so a weaker person (usually a wife) can shoot the weapon also. That low pressure would turn your automatic into a single shot. The only way to get a fresh round in your chamber would be by hand. That could be the worst thing to happen at a most critical time.

MORE RELIABLY, PUMP YOUR OWN

I like pumps because they will feed on all kinds of junker ammo, (the kind I make) whereas autos won't. The modern pump designs and a few automatics allow a shooter to use different quick-change barrels. You can have an 18+" riot barrel, 20" rifle-sighted slug and buckshot barrel (improved cylinder or cylinder bore), a 26" modified choke, or a full choke barrel for ducks and geese. Some home defenders buy a used barrel and cut it off, then squeeze it in a vise. We'll show you how. The same barrel exchanges work with automatics. Besides changing barrel lengths, you can also buy extensions for your magazine tube. More ammo = more versatility.

Because pumps require arm action, they're more difficult to shoot from a prone position. To keep from exposing yourself while racking in the prone---roll to the side, then rack it. LAPD changed over to autos after a big shoot-out with a revolutionary group. Autos are faster, although many people argue they can keep right up in speed. Those who claim that kind of speed probably did what Carlos Hathcock (famous Marine sniper with whom I've spent considerable time) did with his bolt action sniper rifle. Brownell's (gunners' mail order super market) sells lapping compound, and Carlos sat right in his living room working his bolt until it was super smooth. You can do the same with a pump shotgun and make it much quicker.

Don likes adjustable chokes which you can have fitted to any barrel. They allow you to adjust your shot pattern for the load you are shooting (mix 'n match) and also for the range (distance away) you need. While shooting clays, you may want to reach up and adjust with a flip of the wrist to choke the pattern down and deliver a little more effective range. Later, with birds coming toward you, you can turn it open again to get a wider pattern. In defense shooting, you may want to punch through a car with a slug; open to improved cylinder, please. Most full choke barrels cause a squeeze on buckshot which often deforms some pellets and makes them fly erratically. For buckshot, try cylinder and improved cylinder bore choke openings.

SURVIVAL COMBINATIONS

A large part of survival has to do with procuring food, which means hunting or trapping meat.

> MURPHY'S HUNTING LAW
>
> He who takes his rifle into the field will run across birds. If he brings his shotgun, he encounters big game. All the deer in the forest have a wiretap at the State Capitol, and they read newspapers. Thus they know exactly when the season starts and when daylight shooting hours end.

People with my experience will appreciate a shotgun/rifle combination. For the economy-minded survivalist the Savage Model 24F over/under works well. Top barrel can be chambered for the .22 LR, .22 Hornet, .223 Remington or the **.30-30.** The bottom barrel comes in 20 or **12** gauge. Most powerful choice is the 30-30 over 12 gauge. A less-kicking combination would be the .223 over 20 gauge. I carried an over/under for years because I thought one shot should be enough.

> Years ago I was hiking around the logging roads in Oregon and I encountered a good-size black bear sow. I was downwind, so she never found out I was there. All I had was a shotgun and not one of my shells was a slug. I left the bear alone. Had she charged, I would have had to let her get real close before the shotshells I had with me would have done enough damage. Other bears had already gotten close---way too close! Being brave and all, nobody would know that letting a bear get too close would scare me—except the person who did my laundry. I wished I'd had some high power with me.

Springfield Arms offers a survival weapon they call the M6 Scout. Top barrel is either .22 LR, or .22 Hornet, and the bottom barrel is 410 gauge shotgun. The M6 folds for transportation in vehicle or back pack. It has a hinged cheek piece for extra ammo storage. The advantage of any weapon like this is the range it covers. You can shoot birds and small game in close or reach quite a distance with accurate bullet placement.

One thing that peeves me is the law enforcement gun they never made. It contains an electronically sensed bubble device. Once you turn on the switch and pull the trigger, it only fires when the barrel is level. With a duck-billed choke, it would spread a level of shot out sideways. Therefore, any law enforcement officer could squeeze the trigger, point the gun around a corner, and sweep downward from an angle pointing up to clean out a hallway. He wouldn't even have to look, much less aim. The result: Zero exposure for the shooter; max destruction for the guys without a hall pass.

Once you've made your choice in shotgun, it's time to improve it. Very few shotguns on the market today will give maximum performance. To get the max out of any firearm, you have to add "bells and whistles". A little extra money and some of your own leisure time can turn your average shotgun into a super blaster. That's what follows in the next chapter.

II Chapter 2

SHOTGUN IMPROVEMENT

Very few weapons come NIB (New In Box) ready for you to use. If you take your shotgun out of the box and use it the way it comes, you're only using a part of the weapon's potential. Later you'll learn how to make super ammo. So why not shoot it through a super shotgun? Let's improve the gun so it performs for you in the best way possible. After that, you can shoot it with the potential only you know it has.

Your shotgun's stock probably doesn't fit you because stocks are made to fit one universal size---small. The trigger often needs work and the sights need to be improved. Think of a way to carry your shotgun with both hands free; use a sling. How about the recoil? Could you still shoot heavy ammo with a high powder load and reduce the smack on your shoulder? Finally, what about the pattern? Why shoot a round pattern when a horizontally flat one out of a level barrel would guarantee hits on a vertical target? One big complaint we have is this: shotguns are made to be shot during daylight. But shotguns play a large part in defense against crime, most of which occurs in the dark. Therefore, modify your weapon so it delivers your super ammo in daylight or darkness.

MAKE YOUR STOCK FIT YOU

Shooting a shot gun with the wrong size stock will save many birds' lives. We're talking specifically about length of pull, the distance from the trigger to the center of the butt plate. The quick and easy way to tell if your length of pull is correct is to make a right angle at the elbow of your shooting arm, fingers pointed at the sky. Now, set the weapon in the crotch of your elbow. The trigger should be even with the last joint (closest to the end; distal) on your index finger.

If it isn't, cut it down or add length. If the stock is too long, remove the pad and cut the stock; then re-drill the screw holes and re-install the pad. Most often, the stock will be too short and you'll have to build up the butt plate. That's good; you can install more recoil pad. If the pad is sufficient, you'll be installing spacers between the pad and stock.

For hunting shotguns in the North where temperatures change, think about adjustable pads. In summer, you may be wearing only a tee shirt, so you'll need a little extra length. Put one of those ugly recoil pad boots over your stock. You can put more padding between the boot and butt to lengthen it and make it softer. Foot pads designed for the inside of tennis shoes will do fine. In the winter you'll be wearing heavy clothing, so remove the boot.

On a custom stock, think about a cheek piece. Though not commonly found on shotguns, you see many rifles with Monte Carlo cheek pieces because they help align your eye exactly with sights or scope. Likewise, a cheek piece would help a shotgunner keep his cheekbone down on the stock. You could design and make a gel-filled cheek piece for the stock with the same gel used to fill bicycle seats. It would prevent hardwood stocks up against your face from jarring your teeth. To avoid the pain, many shooters raise their heads up off the stock without realizing what they're doing. Raising your cheek up has the same effect as raising the rear sights. That makes the shotgunner shoot over the target.

ADD-ON MAGAZINE CAPACITY

You can buy tube extensions for your magazine and thus carry more ammunition. Careful, though. In hot climate police have left full magazines in cars. In heat the plastic shells have bulged in the magazine so they couldn't feed into the chamber.

SHOTGUN CHOKES

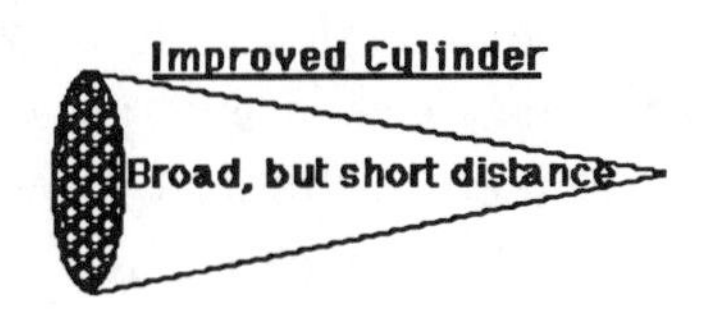

ADJUSTABLE CHOKES

Shotguns used for bird and small game hunting are OK with one choke configuration---modified. To use your shotgun for a variety of purposes, it's best if you can adjust it with a flick of the wrist while you're in the field. That way you can either increase your range, or open your pattern to cover more target area. Finally, if you need a slug for anything, you can crank to improved cylinder and let it fly. Without an adjustable choke, you have to change rings with a tool. By the time you even get started, dinner has flown the coupe.

REDUCING RECOIL

Think about shooting skeet and trap. You reload during the week and then meet at the range. If you yell "Pull" a hundred times or so, you go home with a sore shoulder. More than that, however, you just can't smack yourself with recoil that often without tiring. Therefore, you miss targets.

The most obvious way to reduce recoil is to add recoil pad. Many shooters like compensators, and I'm in that group. Vents out the side of the muzzle divert hot gas to shove the weapon in the opposite direction of recoil. Feels comfortable. Sounds terrible. Mag-Na-Port will cut holes at an angle so the sound goes forward and the recoil is reduced. They call it pro-porting and it costs under a hundred dollars. Just send them your barrel.

Since recoil is the reverse shove of the shot load out the barrel against the weight of the weapon, you can make the weapon heavier so the shot load has more to push against. Thus, your shoulder takes a gentler shove. Drill your stock with an auger bit and insert a lead weight plug or two in the hole. Make sure to leave plenty of side wall in your stock. Cast a lead plug from wheel weights and drop it into the hole; replace the butt plate. The extra weight will make the gun shove as opposed to smack.

HIGH PERFORMANCE FOREARM

A new forearm with a pistol grip will make you faster and more sure. In combat police often get excited and short stroke the slide on a pump shotgun. Result: Empty hull shoved back into the chamber. A pistol grip allows you to work the slide with much more positive action because you don't experience the discomfort (sometimes pain) from the weird angle on your wrist.

To make your own, glue up the kind of (hard) wood you want, then band saw it into rough shape. After that, you'll be rasping and sanding for a long time. Even though you can't install a pistol grip, at least enlarge your forearm. Small forearms allow your fingers to touch the barrel, which can burn after a lot of firing in a combat situation.

ACTION ON YOUR ACTION

Brownell's offers a compound with grit in it to work down the rough spots on your action and make it smooth. Use a marking pen or bluing compound to discover where the rub is. Special oilers are now available which contain the slickest oil made, and that will help your shotgun operate almost friction free.

CARRYING AMMO

Shotgun ammo is bulky. Add bullet loops so you can carry extra shotshells right on your weapon. Stretch cloth bullet loops are cheap and you simply slide them over the butt. Use velcro fasteners pressed together---one on the stock and the other on the cloth---to keep the loops in place. Better than flimsy cloth loops are the plastic ones you bolt on over the receiver. See the Choate catalog.

Shotgun shells will fit backwards into most magazines. Since a lot of defense work with a shotgun occurs when you can't see (at night), be careful to feel the rim of the brass as you shove the plastic end into the magazine. Color code your brass with a highlighter and cut notches in the brass rims of your shells as well. That way you can either see or feel what each round will do. Since most shotguns use tubular magazines, they are LIFO loaders (Last In, First Out).

To load a last-in, first-out magazine (all tubulars) you go to your shotshell supply---perhaps loops on the stock or around your waist. Then you feel the load you need and shove it into the magazine so it cycles next into the chamber.

How will that work? Well, you get into a fire fight with a perp who hides behind a wood fence which 00-Buck can't penetrate. So you reach to the bullet loop and pull a black-marked, single-notched casing out of the loops and shove it into the magazine. The next time you rack your shotgun (LIFO), a rifled slug will chamber so a whole ounce of lead moving along at more than a 1,000 ft. per second will blast a hole through the fence. Fight's over. (Tip: Heat the slug up and drop it into water. Quenched slugs are harder; so they penetrate). [Hotter Tip: There is a way to make a slug explode on contact. Law enforcement agencies and military personnel can have this information for postage & handling only ($1.35). But we won't send it unless you prove who you are.]

ADDING A SLING OR TWO

Not only will your shotgun be easier to carry when slung over your shoulder, but both of your hands will be free so you can use binoculars, tune your radio—-whatever. In addition, your arms won't get tired. You can use a sling in a couple of ways.

Choates makes an add-on sling swivel which bolts over the barrel and tubular magazine on the front. On the side of that is a lug to which you attach an over-the-shoulder strap. (Want a cheap strap? Ask a tennis player who never uses the nylon strap on his racquet cover.)

With two straps, you can wear your shotgun on the shoulder opposite your trigger finger. Now---here's the trick: Put a velcro attachment just about in the middle of the shoulder strap. Attach the other velcro piece to the shoulder of your jacket. With the two velcro tabs engaged, level your shotgun while in a normal shooting stance. To find out where you should place the Velcro, use a carpenter's level on the barrel. With the other tab on your shirt's shoulder (under the epaulet for police officers), level the barrel and mark the strap. Then attach the Velcro. Hot tip: Adjust the length from the velcro position on the straps so you can raise the gun forward and still maintain a level barrel. Thus, you can shoot around corners without exposing much. See *Everybody's Outdoor Survival Guide* on level barrels.

Why would you want your shotgun barrel level? Because you can only miss a target in four directions: Two each---horizontal and vertical. If your weapon is locked by sling from your shoulder, then when it hangs from the shoulder as you walk upright, it can't fire low or high. That eliminates 50% of your misses. With a duckbilled barrel, the horizontal spread makes it difficult to miss left or right.

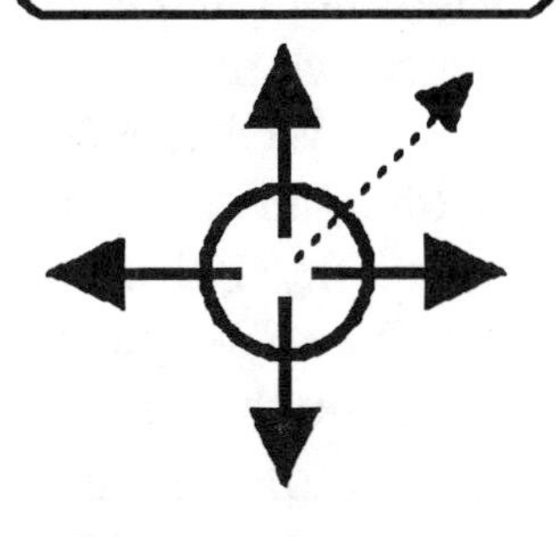

The dotted line is really a combination miss, too much right and too high.

Two slings on your shotgun sounds like a bit much. Still, one would hold the gun level from your shoulder and the other would hang loose. With quick releases, you could undo the shoulder strap and use the sling in the hasty configuration to reduce your wobble on a long shot with a slug.

ACCESSORIES

To keep dust and other clods of dirt from sticking to the oil on your weapon, carry it in a gun case, preferably treated with material to keep the gun from rusting. Otherwise, spray inside the case with a rust preventive. If you carry your shotgun in the field, cloth is better because the noise nylon makes against bushes and tree limbs tells most forest animals that something is out of place. Waterproof gun cases are a good idea, but don't store your weapon in a waterproof bag in an area where the gun will sweat. Moisture trapped in the bag can rust your weapon.

Bullet loops and bandoleers are a good idea. Avoid leather; tannic acid corrodes and eats the brass base of your shells. Since shotshells are bulky, you need a way to carry quantity. Bullet loops on the sling make it sway so you can't hold the weapon still for long shots. For large quantities a carpenter's bag made of nylon around the waist will hold a few boxes. Skeet and trap shooters often carry two bags. One is for empties.

WORKING ON THE TRIGGER

Shotguns don't need the light pull of a pistol (4 lbs.) or a rifle (7 lbs.), but you also don't need a trigger that pulls like a ski jump. If your trigger isn't crisp, have a good gunsmith work on it. While he does that, ask him to cut a slot in the shell follower. Pumps have been known to dump a round past the stop in the magazine (especially with strong, compressed mag springs). So you wind up with a shell between the follower and the bolt which makes the shotgun inoperable. Solution: Your car key through the slot shoves that shell back into the magazine.

Also, carry a back-up handgun. In the engagement business, (law enforcement or military), all it takes is some oil on a primer, a short stroke on your pump, a slip on the magazine stop, and you're out of business permanently. Even a small handgun can save your life.

SHOTGUNS IN DEFENSE. PROBLEM: LENGTH

Like any long gun, a shotgun is unwieldy in tight quarters. So shorten it. Most home gunsmiths simply apply a hacksaw (use a fine-toothed blade) and slice off all excess barrel. Leave some extra (over 18"). If you cut the barrel off too much, you violate an old law (1934) which the ATF will use against you. The barrel can't be shorter than 18 inches and you have to measure that distance from the bolt face with the bolt closed. To be safe, cut the stock short. Several companies make pistol grips and folding stocks for shotguns. If you've chosen a scattergun for home defense (Don applauds), think about the folding stock option. Remember, 27" on a shotgun is required. With a pistol grip and an 18" barrel, ATF will get you.

> Remember ATF, those wonderful people who came to national disgrace on *60 Minutes* after they forced several female agents to prostitute in order to get promoted. Then Waco, using tanks to kill babies; then Weaver, during which they shot his 14-year-old son in the back and his wife in the temple while she was carrying her baby. Would they cheat? Naw... But they will open a double barrel shotgun so the overall length is too short.
>
> **Be careful.**

If you decide to cut your barrel, kiss your pattern ability good-bye. You can re-choke the barrel, but the whole process involves reamers (expensive), dies and threads (over $400 for a kit) and add-ons.

Two obtain a legal but shorter, more usable weapon in tight quarters, and still retain shoulder-aiming benefits, a better way to go is the bullpup. The receiver is set back almost against the buttplate, but the trigger is farther forward. Mossberg makes a bullpup gun, but nobody makes an after-market stock. Still, you can make your own and convert your gun. It isn't hard to fabricate a bullpup stock out of glued up hardwood. The trick is in moving the trigger housing forward. Some recommend a hydraulic hose from a front trigger to activate the rear with a plunger. I don't vote for that because recoil jarring is hard on hydraulic lines. Probably best is two long rods on small U-bolts soldered or welded to make the two triggers Siamese twins with extension rods in between.

DUCK-BILLING. BEST FOR DEFENSE

Situation: You are inside a building in a fire fight. Modify your gun so it never misses. Let's say you have 100 lead pellets to disperse with every trigger pull. You want your message to arrive on target with the best possibility of scoring. Make sure your target feels your influence. Elongating a pattern and crossing it perpendicular to a standing target creates the best chance of scoring. Your area of influence increases on both sides of the target, whereas a simple choked pattern could miss left or right.

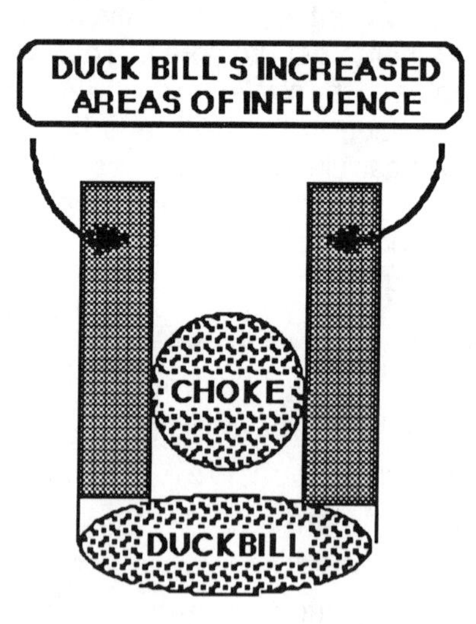

On a sawed-off barrel, duck-bill the muzzle. **Go easy.** Using a couple of 2 X 4 blocks and a vise, clamp down from the top and bottom. Move it slightly; take it out and fire it; check the pattern. Then perhaps squeeze a bit more. **Do not**

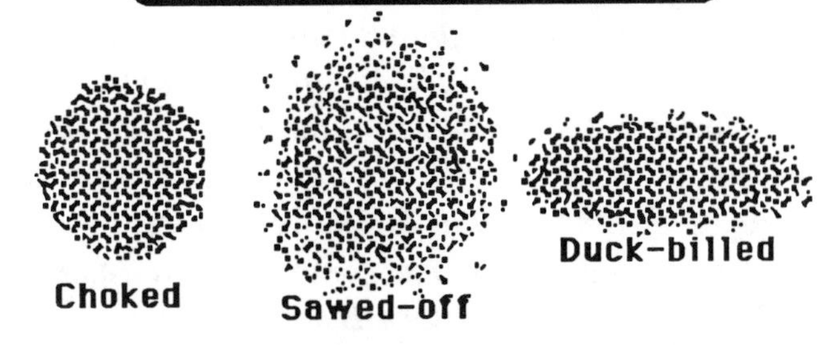

I doubt if anybody shoots a pattern as tight as I drew it choked, but you get the idea. Look what happens when you saw off the barrel---flyers go everywhere. Duck-billing has to be done in careful, squeeze-then-try steps. That's another reason to cut a little over 18". If you squeeze too hard, you can always cut a little more off the barrel not to cut it illegally short.

Be careful.

squeeze so much that you create a geometrically oblong area much less than that of the original circular area. Otherwise, you'll deform shot and pattern.

You can also purchase duckbill, or, horizontal chokes out of police equipment catalogs and silver solder them on your muzzle. Either way, you should get a pattern four feet wide at 20 yards, or eight feet wide at 30. It's the ultimate in self defense shot spreading. To give you the maximum chance of hitting, you want your pattern to cross your target. Therefore, against vehicle, you want a vertical pattern; squeeze the <u>sides</u> of the barrel.

SHOOTING AT NIGHT ON OFFENSE

When I attended the FBI academy as a San Diego

police officer in training, Lt. Brown told us the story of how he subdued a bad perp with his shotgun.

The germ had gone into a pawn shop, asked to see a high-power rifle, then asked for the ammo for the weapon, loaded it and murdered the pawn shop owner. He then began to shoot people in the street. Police came, he barricaded himself, and a fire fight ensued. It was before the days of EAT (Emergency Assault Teams) or SWAT (Special Weapon's Attack Teams). So the cops had a problem., With their pistols against this guy's high-powered rifle—the situation was definitely unsafe enough to make OSHA mad.

> Police departments sometimes engage when the tactical situation is not in the cop's favor. The lure of a high-profile-collar often exceeds the concern for personal safety. This is folly. Take care of number one first. Why risk your life when the courts are simply going to release the guy early anyway?

It was one of the very first crimes of its kind, and after a series of events, the department decided to send Sgt. Brown in after the guy. What weapon did he take? Shotgun; smart guy. So—Brown goes upstairs, carefully from room to room in the dark. Suddenly, a door flies open. Brown shoots too quickly. Why? The perp hid in a closet, and after having been in a totally dark area for a while, his eyes had adjusted completely. He had night vision. Brown came into the building from daylight outside, so his eyes hadn't adjusted. The gunman turns as Brown shoots. Immediately, Brown shoots again. Result: An armless germ goes to prison. In the rush to shoot, Brown hadn't been able to line up on target, and both rounds had each hit the perp's shoulders.

All shotguns are made for daylight shooting. Most military engagements and criminal encounters occur at night. It's up to you to adapt. Feel your shotshells and the places on your gun where you know your barrel will be level. Listen to incoming and determine location. Learn to look quickly left and right of your target so your "night purple" helps you see your target.

ADDING SIGHTS

You ought to add buckhorn sights to your shotgun so you know for sure exactly how it shoots. Your shotgun is only doing half the work it could be doing if you don't use slugs and sabots, both of which require sights. When you do install those sights, think about night shooting. Install tritium dots if you can afford them. Otherwise, apply Tulip Paint (a glow-in-the-dark paint from fabric stores) you can charge with a light. Either way you can see your sights at night. I don't like lasers for a shotgun because they cause you to pinpoint your target and thus suffer lag time before shooting. Also, you can't use them in daylight. (When I co-authored ***SECURE FROM CRIME*** with high-ticket bodyguard Craig Huber, he asked me this question.) "Do you really want to trust your life to a battery?"

To me, a ventilated rib is part of a shotgun and it wouldn't be complete without one. On top of your ventilated rib, drill and tap for two dots. When my two dots are aligned, they create a point of aim in the center of my shot pattern. Find out if your weapon shoots the same way. Pattern it against a giant cardboard box from a distance.

Scopes also make a shotgun easier to shoot, but they restrict the field of view. Since most slugs shoot a broad two minutes of angle or more, a scope won't help a whole lot because: A.) The weapon can't place the projectile on the point of aim anyway. B.) A shotgun only shoots a bit over a hundred yards and you don't need optics to see that far. C.) The thing gets in your way if you have to quick-shoot, which is primarily what a shotgun is designed to do.

OTHER SIGHTING SYSTEMS

If you use a shotgun in the assault mode, you may think you have to "stand up and fight like a man." Many men who have held to that code are now dead. As a former Green Beret parachutist, cop, hero, whatever, let me tell you: "Stand up" is what you do when you're a comedian. In gun fights, it gives the other side a chance to kill you. So duck, run, crawl on your belly like a snake. Above all—**don't risk getting shot!**

In combat, I'm in favor of zero exposure. That means you keep your body free from incoming. Three ways to do that are: A.) Not to engage, period. B.) To get behind safe cover before engaging. C.) To shoot first. In pitch black with no light available, the tactical lighting situation may be in the bad guys favor. Like people inside a theater during the day, you can't see them, but they can see you perfectly. Why? They've been there in the dark long enough for their eyes to adjust. The problem is severe if you rely on eyesight alone, which is what most police do.

AIM WITH YOUR EARS

This is better: Rig your shotgun to fire by sound. Conical mikes are available today which can be turned up so you can hear a man breathing. Instead of using a flashlight or red dot laser sight on your shotgun, both of which disclose your position, use sound. Simply attach a conical mike to the underside of your barrel. Rig it so it turns off when you pull the trigger. Turn up the volume on one earphone. Sweep the room. When you hear your target loudest, squeeze the trigger. Of course, you risk friendly fire problems during military and law enforcement sweeps when you do this. The solution is to adapt the MILES system now being used by the military (probably, Military Identification Laser Engagement System). Build an inexpensive short-range transmitter which when received by the shotgunner's small directional antenna, turns on a faint light on the shotgun's receiver which only the gunner can see.

Systems for shooting in the dark could involve flashlights. Choate in Arkansas sells the Tac-Star light which attaches to your shotgun and hits a perp with high intensity light. Stunning. Now you have a perp illuminated and you can pull your shotgun trigger immediately if need be. What they don't tell you is this: In a high intensity gunfight, you may hit the light switch by accident. Worse, the perp knows his business. He tosses something in the wrong direction, and you shine the light at the noise. When your light goes on, he kills you.

MAINTENANCE

After lapping the rubbing parts, use Pro-tec grease or oil to make them slick. Not too much. A lot of oil in the action of a loaded gun can deaden a primer. Learn how to clean your weapon. While cleaning, however, check a few things. I want **no** dents in the magazine tube. Otherwise you may get a failure to feed. Use a magnifying glass in strong sunlight to check the receiver for cracks. Place a small coin on your bolt face and close the action slowly while the barrel points straight up. Snap trigger; listen. The coin (dime) will fly upward and land again. If not, you broke a firing pin.

Just like the shells you can buy at the store, your shotgun has limitations. But improve it. Make it fast and accurate. Make it spread shot or shoot farther. Make it work for you both night and day. Once you make the improvements, its performance will amaze you. You'll also make some shots people won't believe.

Build superior ammunition to shoot through an improved weapon and gain superior performance. Recording any shot load's performance with that shell's loading data is most important in achieving best results. That's why we devised this invaluable tool.

Loading data goes on the top section, and performance goes on the bottom. Enlarge the form for records; reduce for labels to go on storage containers.

SHOTSHELL RELOADING AND PERFORMANCE RECORD

LOAD DATA

GAUGE SHOT SIZE SHOT WEIGHT VELOCITY RANGE

POWDER POWDER WEIGHT OZ/POWDER PRESSURE PRIMER CASE

PERFORMANCE

SHOOTING DISTANCE CHOKE SETTING BARREL LENGTH TARGET

Number of shot, this load

% in 30" % out

SCORES: Skeet Trap Clays Game

RECOIL
Heavy
Medium
Light

COMMENTS

II-Chapter 3

SHOTGUN USE

Begin with your shotgun by determining how far you can effectively place a load of shot. If you bird hunt, learn how far your shot load will travel and still effectively bring home dinner. Shoot a pattern with the load of your choice at increasing distance to make sure. As you do that, you might increase the charges of powder and shot you drop in each shell as long as you stay within the limits specified in your loading manual. Well and good, but once you establish the limit don't exceed it. You'll only wound birds and the pressure you'll create in your shotgun from hot loads isn't worth it.

PATTERN YOUR SHOTGUN

This is fun and afterwards you'll gain a lot more confidence in your shotgun. With the use of slugs and sabots pattern is probably the wrong word. The idea is to know exactly where your gun shoots. You don't have to sandbag your weapon, but you must pay attention and use rifle shooting techniques so you eliminate shooter error. Either use sights or align the beads on your barrel properly with the target to get accurate results.

Aim at a stationary aiming point. Either from a good sitting position or perhaps by using a rest, make sure the weapon is steady, i.e., wobble free. When the sights are perfectly aligned with the target, squeeze the trigger to make the round fire as a surprise. Now, note the results. First, find the center of the pattern. Don't use flyers (odd pieces of shot or rock in your load which might have flown off outside the target area). Make diagonals at right angles through the shot pattern to find its center. If the center of your shot pattern is not on target after double checking, you'd better see a gunsmith.

Find out how many buckshot pellets will hit inside a 12" circle from a distance of 20 and 30 yards. You should have at least 50% inside the 12" circle. For testing bird shot use a 30" circle at 40 yards. Learn how your shotgun patterns with different shot sizes. Also, learn where it shoots with slugs and sabots. Mix 'n match loading may present some difficulties with pistols and rifles (point-of-aim changes, confusion, different recoil, etc.), but with shotguns it is definitely the way to go. You want your shotgun to be effective at all kinds of distance, through barricades, inside buildings at short range, etc. To do that best, you have to load a variety of shells and know how they'll shoot and where they'll land.

HOME INVASION DEFENSE TACTICS

This is about defending your home and person against attack while you are in the home. A shotgun outranks a handgun in firepower. Since you can learn to shoot one more easily than a handgun, the job of home defense belongs to the shotgun. Before you blast, however, you need to know about the situation in your home. Burglar? Friend? Child? Gather intelligence.

If you budget for home defense every month, the first time you collect under $50 you can install an intercom system in your home which will gather sound intel from any one of four locations. It also provides you with an opportunity to communicate with anyone outside your bedroom. Another good idea is to employ a full-time bodyguard—-the four-footed canine variety. Early warnings are often enough to get you up and in good defense position. Even better, dogs often scare off would-be attackers.

When someone invades your home while you are present, you have a dangerous situation on your hands. A high percentage of these crimes turn violent. Since you can't foretell whether an intruder only wants goods or is half crazy and looking for cheap "helter skelter" thrills, assume the worst---and be careful! Two crime scenarios are currently employed by home invaders. If it's only a burglar, he may not bother you if you don't bother him, and many home residents have a strategy built around that theory. Even if a psycho is in your house, a safe strategic plan involves shutting you and your loved ones inside your sanctuary and letting him take what he wants. You may have a long wait. Generally, he will cut the phone lines before he enters. To get police help, use a cellular phone in your safe room (also called sanctuary). Preset the number 911 so all you have to do is push "enter" in an emergency. If you call for help, get a reasonable estimate on arrival time, and describe yourself over the phone so you don't get mistaken for a burglar and shot by the police you called!

Don't make deals. Trust me on this; when I wrote ***SECURE FROM CRIME***, I was amazed at the mentality I found in the streets. These people are almost compelled to kill you. They can't back down without losing face in the only supportive sub-society they know.

The latest crime scenario is HIR (Home Invasion Robbery), in which multiple perps, generally from the same gang, ram your door and enter en masse. They terrorize the occupants until they've learned where all the valuables are. Rape sometimes occurs. Many victims feel so glad to be alive they have no care whatsoever for their property losses.

SHOTGUN IN DEFENSE

The average person half awake, frightened and under stress, can make some big mistakes, and shotguns make big holes in things. Also, a long shotgun or rifle is an open invitation for a gun grabber if you let a perp get close to you. Some prefer a pistol for home defense because you can operate it with one hand. Don prefers the shotgun because every street thug knows what Doc Holliday taught at the OK corral: Shotguns out-perform pistols unless a magnum handgunner is outside of the shotgunner's range, such as over 150 yards. Most people who invade your home carry a handgun because they need to conceal. In close quarters, they know the rack of a shotgun puts them at a distinct disadvantage and means real trouble.

Of course, a large size of shot, such as 00-Buck will go through a wall. Solve that problem by using flechette loads in shotshells which penetrate burglars, not walls. Also, a pistol grip on a shotgun for close-in work is best. You can order the grips from all kinds of mail order stores or gun shops. Unbolt the big stock, install the small. If you ever want to use your shotgun as a shoulder weapon again, reverse the process. Choate makes a stock you can fold up next to the gun's action. The pistol grip is then usable and the gun operates in close quarters. As soon as the shotgunner needs to shoulder the weapon, he merely folds the stock back.

I assume that you've at least gone to the library and borrowed a copy of ***SECURE FROM CRIME, How To Be Your Own Bodyguard.*** In the event they don't have a copy, you can ask the acquisitions librarian to order one;

the ISBN # is 0-938263-18-8. In that book, we show you how to convert an ordinary bedroom into a sanctuary. One of the elements is a solid core door.

During a home invasion, never try to be a hero. **Don't go on offense!** People get shot or captured inside their homes when they move about. Any intruder can simply listen and wait until the homeowner walks into a trap, turns on a light or shines a flashlight around which gives away his position. The exception to the above rule occurs if you have to block an access route to your children or loved ones. Otherwise, stick with a defensive operation. If you have a solid core door on your bedroom with a firing port (see ***SECURE FROM CRIME***) barricade yourself inside. Remember, walls don't stop bullets or shot. Drill a hole and fill the space with sand. Inside your sanctuary, take up a position behind your mattress, and simply wait things out. If your intruder tries to come through the door, provide the appropriate welcome.

Plan the lighting so the tactical situation is in your favor. Two rules on tactical lighting advantage are:

A. Always shoot from the dark into the light.
B. Place yourself so your enemy will be a silhouette from light behind him.

Using the first rule, your enemy can't see you (very well) but you can see him. Remember, muzzle flash gives you away. Shotguns normally don't miss, but don't count on it. Fire; then move immediately to another position with cover. Using rule two, you line up so the passage way toward you is in front of a window. That's common at dusk or dawn, where lighting inside is poor, but anyone moving in front of an illuminated background loses.

You also need to have the noise level in your favor. The rule:

Always make <u>less</u> noise than your enemy.

So you move in silence, but he has to make noise on the stairs. If you have the time, you can crawl on your hands and knees to distribute your weight evenly. But when he takes a step, you can hear and locate him. Leave the floorboards and doors in your home squeaky. Learn which location produces which sound.

PRE-DEFINE YOUR FIELD OF FIRE

Especially with more than one family member shooting, you must set up safe, no-fire zones. These are places toward which nobody will fire. That way, you don't risk the chance of shooting each other or children sleeping in a separate room. Don't fire in their direction.

BORE SIGHTING

I recently did a radio show for a host back East who told me afterward he was blind. His question was: "How can a blind person defend himself in his home?" Not many people know this, but you can be as good with a shotgun as anybody if you're blind; besides, shooting without sighting is a lot safer. Perhaps this defense is a little rude and not politically correct in our current anti-gun atmosphere. But my opinion is: Anybody who invades your home shouldn't get a break.

If you could shoot your shotgun at a perpetrator without having to sight over the weapon, you would be at zero risk. I like that. This is what you can do. Pick out an ambush spot in your home. Drill a hole the size of your shotgun barrel in the door down low. When the shotgun muzzle rests in the hole and the butt is on the floor, the line of the barrel will connect with the perpetrator standing outside your bedroom.

With the door open, stretch a piece of string or building line down to the floor at a point where the butt of

DRILLING A HOLE THROUGH YOUR SAFE ROOM DOOR FOR FIRING YOUR SHOTGUN

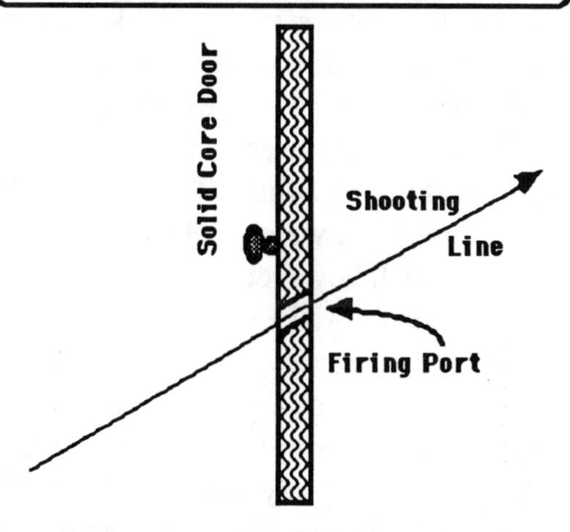

> Your shotgun can shoot through it upwards to take out anyone standing in a pre-determined spot. For example, a perp standing on a pressure switch at the head of the stairs would make an excellent target. Without the pressure switch, you'll need some kind of visual system---perhaps a peep hole mounted in the thickest part of the door.

your shotgun would go. Mark that spot. Determine the angle of the string. Drill the hole on that angle. With your shotgun action open and the magazine empty for safety, place the muzzle in the door's new firing port (aimed upward and out) and the butt of the weapon in the place you marked. Test the line of fire by running the string from the action out the barrel to the ambush spot.

Place a pressure switch under the carpet at the right location. If ever attacked, close and lock your sanctuary door, then place your loaded shotgun in position. When the pressure switch activates a signal, you can pull (gently) the trigger with a string and be completely out of harm's way as your shotgun blasts. This is one occasion for an automatic shotgun.

IF YOU HAVE TO SHOOT AT NIGHT

Most crime occurs in the dark. Many home defenders aren't trained for night combat. Those who shoot from the hip leave both eyes open as they do so. Bad tactics. With the weapon in your hands, make sure to close one eye tightly upon firing. The open eye's iris will close down and lose night vision. The eye you closed will still be able to see well, however, so you can continue.

HITS AND MISSES

Some people believe you just can't miss with a shotgun. That just isn't true. It happens—-often at close range because the shot needs distance before it can spread out. The idea of shooting any gun is to hit, which is the opposite of miss. Therefore, if you examine all the ways you can miss and eliminate them, you'll be ready to shoot with Wyatt Earp. This is easy to learn, because you can only miss your target in one of four ways.

You eliminate left or right misses merely by aligning the barrel with the target. Besides that, moving targets require you to follow through. Don't quit swinging after you pull the trigger; keep swinging along with the target after the gun goes off. Otherwise, you'll shoot behind most moving targets.

Now let's deal with low and high. Low misses on human targets frequently hit anyway because the round projectiles bounce off the ground surface and come back up into the target. When all is said and done, the one miss you have to worry about is high. Why do we all shoot high?

If you get a chance to shoulder your shotgun, you can look down your ventilated rib, line up both dots, place the target's center of mass just above those dots and squeeze the trigger. Shooting opportunities like that only happen in dreams. What really happens is the clay pigeon flies, you raise the weapon, try to track it and yank on the trigger. In combat the other guy is armed (or you wouldn't be shooting, would you?) so you throw the weapon up there and concentrate on the target. Either way you commit the major shooter's sin: **When you lift your head high off the stock, it's as if you raised your rear sight.** Result: You shoot high—-over the target. It's important to keep your barrel level in combat. That's why we came up with the shoulder strap adjusted with velcro tabs while in your shooting stance.

HIP SHOOTING

In survival shooting, you don't often get a chance to shoulder the weapon before you start popping primers. Perhaps a nearby snake jumps out at you. Rabbits and ground squirrels run for cover quickly and you'll have to shoot fast. Finally, you may be dealing with some human germs who will shoot first if you don't. Either way, you have to learn to shoot with a level barrel as I taught in ***EVERYBODY'S OUTDOOR SURVIVAL GUIDE.***

Without a shoulder strap, this is the basic method: Put your hand on the pistol grip in the same place every time you hip shoot. Dig your thumb into your hip in exactly the same spot every time. After that, with your non-trigger hand, find the place on the forearm (either forward or back) where you hold to make the weapon's barrel absolutely level. Use a carpenter's level on the top of the barrel to make sure, and then mark the place on the forearm where your (left) hand fits. Either use tape you can feel, or glue some plastic feelers (cut a plastic fork's tines off and shape them with a file). When your hand comes to rest on this spot, and your (right) hand is positioned at your hip, all you have to do when shooting is keep your back straight. Like a human gun platform, you can pivot from the hips and take out several targets moving in all directions. In your own shooting stance you'll know the barrel is level; all you have to do is line the barrel up with the target and let fly.

Another way to achieve barrel/target line-up is to raise the weapon up under your shoulder and visually point the barrel at your target. Soften your focus and allow your peripheral vision to take in both the barrel direction and the perp. With your arm clamped down around the stock, you absorb recoil easily. Both Don and Dave prefer this if you are not standing on level ground. As most of us know, fire fights are fast, fluid and filled with stress. Practice. (Inexpensive to do if you reload; it will only cost .08 cents per round.) Once you find the way that works for you best, stick with it. Now, practice more.

If you get a chance to shoulder your shotgun, don't assume the shot can't miss because the spread will make up for your errors. Instead, concentrate on one small, vital part of your target and try to place the shot there as if you only were shooting one slug. Don't just throw shot. Rather than shoot in the general direction of your target, be specific.

MIX 'N MATCH LOADING

To become the ultimate combat shotgunner, load your magazine to take advantage of the shotgun's versatility. Fill your magazine so you get high-energy shot spread, slug whump, and sabot range. Assuming you have a pump or automatic, your weapon feeds from a tubular magazine. Shotguns with tubular magazines are LIFO loaders---Last In, First Out---which refers to the order in which different shells fire. Here's what's best for survival (not hunting). Pull the (wooden) space plug out of the magazine. Load in this order: Sabot, slug, 00 buck, 4's, 6's, and perhaps 8's. Now, when it comes time to shoot, they go out of the barrel in reverse order; the 8's fire first and the slug and sabot fire last. The question is: How will you react in a fire fight when you know which shell will produce exactly the result you need?

If you've practiced a lot, you know which number in your magazine does what job. To get to the load you need, you can use one of two methods: Either rack your weapon to eject through, or shoot through until you get to the load you need. Racking, or shucking, is easy to do if you have a pump because you don't have to take the weapon off your shoulder. Shucking rounds out of an automatic presents more of a problem, so you'll probably choose to shoot through, which is the method most shooters (pump or auto) prefer anyway. Remember, however, that shooting usually discloses your position, especially at night---when most firefights occur. By the time you work your way down to the round you need, muzzle flash will broadcast your position to the whole neighborhood.

Dave writes: To think you can unload one kind of round and then reload with another is unrealistic. With hot lead flying and adrenaline pumping, all your attention should be focused on tactical shooting, not fumbling around changing ammo.

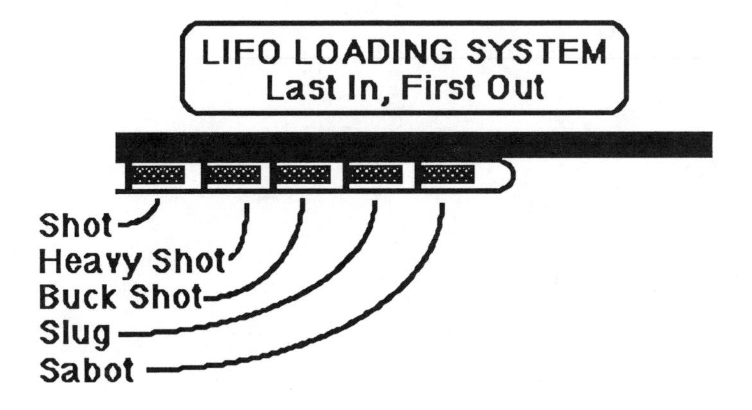

You can't cover a lot of distance with a shotgun, but you can be sure the distance you do cover will feel your influence. Shoot within the range of the load you have in the chamber. Set your choke for the job you need, and use the super ammo we'll show you how to build. When you do all of that, you'll be most effective.

II-Chapter 4

RELOADING SAFETY FOR SHOTGUNS AND RIFLES

Think of reloading as the opposite of shooting. At the factory where they made your original round, let's suppose that the bullet was second class, the powder off a grain or so, and the primer hole plugged. Obviously, that cartridge would not shoot like any other.

When you reload (unshoot), you will replace the bullet with a new, high-quality projectile. It will not only weigh exactly the same as others; it will be concentric so it spins true, and it will perform perfectly during the terminal ballistic phase. Your powder measure will be perfect, because you pay attention to the loads and weigh them often to check during the charging process. Your case will be cleaned and trimmed. The flash hole for the primer will have no obstruction.

It's important to keep accurate records for every group of rounds you build. Use our forms. We release the copyright so you can make 50 of them into a spiral bound book (at Kinko's or any print shop). Normally, you'll build rifle rounds in five-shot lots. The records you keep in the special rifle book we designed tie the bullet performance to the powder load, primer, case and bullet. Color code each lot so it matches the color in your reloading book.

In my early reloading days, I didn't do this. Then when I got terrific performance out of a certain box of cartridges, I couldn't duplicate them. Also, some rounds I put together were terrible. Duds meant I had oil on my fingers that day when I loaded primers. Without a date on that lot, I couldn't locate "brother" rounds to be discarded or shot in practice. I never knew why things went bad and later I built other bullet/powder combinations that produced similar, lousy results. **Keep detailed records!**

If you get a round that surprises you or is unusual, stop. Don't shoot any more from that lot (bunch of rounds you put together) until you discover why. Usually, that means going back to the bench and pulling the rounds apart. (Are you sure what kind of powder you used? If not, incinerate it.) Make sure your powder measure is empty before adding different powder. Powder mixing can be **BIG** trouble.

Want to save a few cents by reusing a damaged case? A new eye is priceless. Examine every case and hull you will reload. Damage of any kind? Chuck it. That also goes for any powder or primer about which you are not sure. You **must know** the exact powder and primer you're using. Otherwise, throw it away. Putting away everything except one container of powder, one box of primers and one box of projectiles costs time; failing to do that can cost a new gun.

Go ahead and drink alcohol while you reload. The buzz will make the time fly by. If you drink, you might smoke too. Smoking may set off an explosion, but if you drink lots of liquor you won't feel the concussion. You don't have to smoke to get a good fire going, however. Use a vacuum cleaner to scoop up spilled powder. When the vacuum cleaner sparks, so will the whole area.

Sorry—-no children in the reloading area or room unless supervised and quiet. You can't afford distractions. You'll need to concentrate on precise & repetitive routines. Children do things like mix 150 grain bullets in with 165's. It's also fun to put one of those little pieces of shot down in a primer cup. They may not have your attention now, but the first time you place a primer in the cup over one little #8 shot they will. The rule is: Only highly competent adult shooters get to place their hands anywhere near the presses or components.

Keep component storage at correct temperature and humidity ranges. Airtight is the general rule for reloading components. Even brass and copper suffer some from oxidation. With just a little care, you can store both components and loaded ammo for many years.

Don't take reloading data from any source other than a reloading book or the manufacturer of the components. Then pay close attention. Don't experiment! The experimental approach to reloading can be full of surprises, some disastrous.

Weapons have design limits. You can't make your shotgun shoot high velocity, heavy shot loads without damaging it. Sometimes though, you also destroy the shooter. Reload **only** for performance within the limitations of the gun.

Do not reload in a hurry. When reloading, pretend you're a government employee. Go slowly; be deliberate. You know how bureaucrats don't need God because they've found the great eternal resting place? Do likewise. Take your own sweet time. If you haven't slept well the night before, don't reload.

Measure twice; live once. Don't think weighing the same powder charge twice is dumb; it can be a life-saver. If you ever load anything close to a maximum load, hand weigh each charge. Develop a list of procedures you always follow. Example: Check book, double check bullet weight for this charge, weigh powder. Set powder measure. Re-weigh powder charge. OK? Now, run twenty cases under the measure. Watch to make sure no (humid) powder sticks in the drop tube. Visually inspect each powder-loaded case with flashlight to make sure all levels of powder are the same. Stop. Throw another powder charge and weigh it. Has the powder measure changed on you?

Let me tell you about myself. I've written 8 books; I'm creative—-not neat. If I had to run a singles ad it would say, "Do you like garage sales and flea markets? Marry this nice man who is 50% off." I was a very creative warrior as a Green Beret. I never had a regular job and I seldom do the same thing the same way twice. I've lived everywhere, speak several languages, traveled all over. However, when I reload, I try to have the personality of a dial tone. Safe routines. Over and over. No distractions. **No experiments**. Here's a quotation from the Book of Life: "Go thou and do likewise."

Now—-I follow all the rules—-not because I am cowardly, not because I have a lot of time, and certainly not because I am by nature a methodical or deliberate person. Also, I'm not afraid of dying. But I don't want to die stupidly. I don't want my gravestone to read:

Here lies Don, Green Beret at large
Died like a dummy from---
a double powder charge.

Several guns have been ruined and a few shooters hurt because they tried to make the weapon do more than it was designed to do. Shotgunners will always be tempted to load more shot and more powder so they can reach out and touch a duck at longer distances. Terrible idea. The same goes for pistol and rifle shooters who want more range or more wallop. Why should you shorten barrel life and risk damaging a good weapon just to achieve a little more?

Bullets don't go any farther or faster when you reach one of two limits: A.) Powder capacity in the case. B.) Pressure in the chamber. If you come close to exceeding either one of these limits, you create danger. Fill it to the brim may work for coffee, but with powder you may be filling it, then compressing it with a bullet. That ought to wake you up. The same kind of surprise awaits you when you ignore pressure signs.

Instead of making the gun do a little more, you do more. Stalk better and get closer to what you have to shoot. The practice will be good for you. Work on your neutral scent, noise discipline and camouflage to get game to come in closer to you. If you find that you absolutely need a longer shot, **don't try to cure the problem with a heavier, high pressure-load!** Buy a heavier gun.

For purposes of this book, we want you to learn to

reload, with no danger to yourself or teammates. Therefore, start with a load velocity and bullet weight at the low end of the range. Keep improving your loading techniques and test your rounds until you know for sure what every weapon does. Once you build five rounds that group tight, shoot flat and provide good energy at the target distance, use your new rifle record book to create superb repeat performance.

RIFLE CARTRIDGE: LOADING & SHOOTING DATA

a
date time
rifle caliber bullet type shooting range
b
mid-range powder primer case
m
c
humidity temperature elevation ASL
d
load velocity muzzle Nomographic: midrange 1/4 energy at target

GROUP ADJUSTED?
\+ −
elev windage

	elev	windage
A		
B		
C		
D		

h

COMMENTS

Wind Values Assigned

Corrected to:

WIND direction velocity

k

From the book, *AMMO FOREVER, The Complete What To Shoot and How Manual* by Path Finder Publications, 1296 E. Gibson Rd, E-301, Woodland Ca

This form does more than create a record of rifle loading data and subsequent performance. Use it for labeling. Shrink the form and glue it to your cartridge boxes. Once you build a set of shotshells or rifle cartridges that perform well, you need to identify the boxes of shells. Later, you may choose a few shells out of each box to mix 'n match for maximum shooting efficiency.

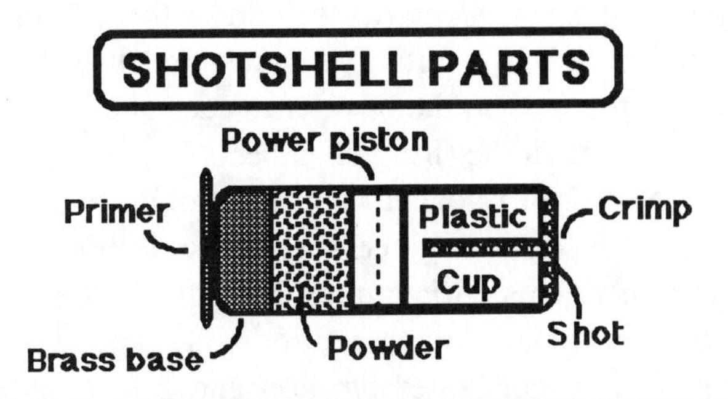

Shotshell reloading is rather simple. You can buy a variety of books that take you from A to Z. Because Path Finder only wants new and unique material, we don't cover the reloading process in step-by-step fashion.

II-Chapter 5

SHOTSHELL RELOADING

Think about reloading. It's really un-shooting. You put everything back you used. A new primer will ignite new powder to push a new wad which contains a new load of shot. When everything is replaced, you crimp the top and it's ready to "un-reload," or shoot again.

Shotshell reloading is simple, easy, and is almost error proof. Many press manufacturers tell you not to worry about component substitution unless you're a purist (which I am). Furthermore, the instructions they furnish with the press are complete in a couple pages. Resize, prime, drop powder, insert wad, drop shot, start crimp and finish crimp---done. In my opinion, the most important factor in making a quality reload is the crimp. Why? Because the crimp puts even pressure on the shot and therefore the wad so the powder in each shell develops the same pressure in the firing chamber.

Shotshell reloaders get into trouble when they try to become ballistic engineers. They re-design the press or make adjustments. Sometimes they substitute components and build a shell no manufacturer would try. By changing primers, they get better ignition and a few more fps in velocity, but raise pressures dangerously. They want more shotshells per hour, so they install automatic feeding tubes for primers, which build up static electricity and ignite---all at once---because they failed to ground the tube. Then the ultimate error: They don't check volume charges by weighing them. Most shotshell reloaders use a charge bar (volume) to drop powder. By inserting a bushing into the charge bar to take up space, they reduce the load. But some bushings are mismarked. Result: Wrong powder charge.

A number of shotgun reloading manuals contain all the data you'll need. Powder manufacturers supply manuals free. Use recommended loads. Otherwise, you may invalidate the warranty—-on your head. Pay close attention to what you're doing. Just weigh the shot charge. Don't use a hot primer, too much powder and a heavy shot charge to blow the gun apart.

Now, after we make you paranoid, we tell you this: Your shotgun offers versatility. Why? You can make all kinds of shells shoot various ranges with different projectiles and powder. Of course, don't think you can change powders and use the same charge bar. Different powders burn either faster or slower, with more pressure or less in the chamber. If you use the same charge bar volume with a faster burning powder, you're in for a surprise. Also, some powders are more dense than others. So the same volume on the charge bar weighs more. Since the weight of the powder is equal to the boom of the blast, you could cause an explosion. To stay safe, "remain my kind of reloader; go by the book." Also, don't trust your charge bar. Weigh the charges to make sure the bar delivers the right amount. If your bar or the sleeve you insert doesn't drop enough powder use a round file to open it up and increase the volume. On the other hand, when the hole is too big, replace it. Get a new charge bar or bushing.

Changing shot size is not so severe. Sure, a heavier charge will put more pressure on the chamber. But you can fix the problem simply by weighing the shot charge after you drop some through the bar. The amount of shot with which you load depends on what shot pattern density you require and how much recoil you can stand. Loads are for lead shot (not steel shot) and start with 1 oz. (437.5 grains) of shot and go up to 1 & 1/2 oz. of shot in the 12 gauge (2" & 3/4" case). The way your shotgun patterns with different loads of shot and powder will determine what's best.

A more common problem occurs with shot volume, something you can fix with a wad change. If you don't match your charge bar to the size shot you're using, you'll use up more volume in the shotshell so the final crimp won't be right. One reloader I interviewed changed from 7's to 8's, which caused some vacant space in the shotshell volume. When his star

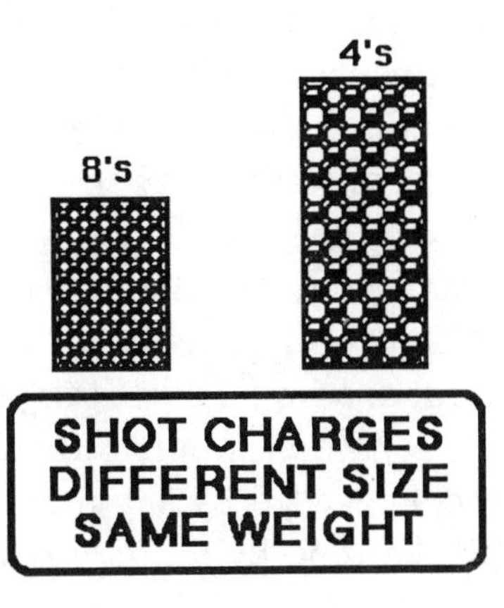

crimp didn't close all the way, he dropped some candle wax over the hole in the middle. Note: The scale you need for shotshell reloading must have a capacity to measure 1,000 gr. to weigh (heavier) shot.

Building exotic loads with explosives in the shot compartment is excluded here. You can write for the information if you identify yourself as law enforcement or military. Explosive shotshells can open a water tank and cause floods, detonate fuel tanks, etc. H.E. rounds take out personnel. You can buy shotshells that deliver concussion and flash loads. In a dark room, for example, a flash load destroys the night vision of the perp. That makes police entry relatively safe.

PRESS CHOICES

I use the Lee Load-All; with dies it costs less than $50. This is my choice for single phase press for custom loads because I activate the shot-drop by hand. One reloader I know, for example, mixes oatmeal with his shotloads because he makes special (low recoil) ammunition for his wife to shoot out of a 12 gauge for home defense. I suppose she'll ask the intruder, "Had breakfast, sir?"

Some reloaders get full energy on target and increase wound potential when they cable or tie shot together. Simply drill small holes through the soft lead of 00 Buck and make a necklace. Solder the balls in place to contain the pattern at a certain interval. One large ball soldered to the next size smaller will make the two twirl through the air. The power of the forward moving shot drags the tight cable in between so it slices through a lot of material. This is not a good jungle or dense woods load; one twig slows down and perhaps tumbles it.

Remember, substituting components is not recommended by any manufacturer. It's done in the field all the time by military operatives. Generally, however, you stay with the shot charges by weight and the powder and primer recommended by the manufacturers. For example, don't use a primer with an open (not covered by plastic or paper) hole if you choose a ball powder. A few of those little powder balls will fall down into the primer and make a bang you can't imagine. If you drop some lead and add #10 wire (large paper clip wire) to make a flechette round, you weigh the charge and adjust the powder accordingly. If you happen to be in a situation where you've used all your slugs and you think you'll need them, pry open any field load, dump the shot, and dribble it back into the cup as you add candle wax or melted paraffin (household canning) wax. Epoxy costs more.

Besides my Lee, I've used MECs (Mayville Engineering Co.) progressive turret presses for skeet shooting. Progressive presses turn out a good number of shells per hour with little effort. All you do is simply rotate the turret and every stroke of the handle produces a finished shell. At the same time, the press performs functions at other stations. It reprimes a newly sized case, drops powder, pressure-places wad, drops shot, begins a crimp, etc., all at the same time.

Most progressive presses are just about error proof. In the powder-drop phase, for example, a sliding bar moves across the drop tube. The hole in the bar determines the powder charge by volume. (Once again, don't trust it. Weigh a charge to make sure.) You rotate a turret or progressive press every time an operation is completed. With an operation like that, it's difficult to make a mistake. However, progressives make it difficult to build a custom load.

COMPONENTS---MAKING GOOD CHOICES

As with rifle and pistol cases, shotgun cases have to be clean and kept cool and dry. Unlike rifle and pistol cases, shot-gun cases “MUST” be separated as to brand

and type. You'll find a great difference between straight and tapered inside case walls and, therefore, loading components. Wad pressures have to be changed to accommodate case capacity. Internal differences cause the same powder and shot load to take up too much space, so perhaps the shell won't crimp. Keep your cases separate. While separating cases discard any case with cracks, holes, tears, loose head-to-body fit.

TWO KINDS OF SHOTSHELLS

Shotguns are for hitting targets perhaps 40-50 yards away. You get more range when you use larger shot and full choking. Small shot suffers from what I call "energy decay." Even though the charge flies out of the muzzle at a hefty velocity, each small ball doesn't have enough weight pushing forward against air resistance, so it slows down substantially. Extra powder doesn't do much for effective range because you're dealing with spherical (round) ejecta (the stuff flying out of the muzzle).

You have to practice. Typically, you either go to a range or you buy a case of birds and a thrower. Then you take turns---one guy throws and the other shoots. For that kind of shooting you can load 7's, 8's or 9's, and blast away. If you do a lot of shooting, you probably would use a progressive reloader to produce shells in which you would drop a medium load (ounce and a quarter) with a cheap wad.

Hunting loads and defense loads require a little more thought and time to build. You can (and should) cast your own vaned lead slugs in a Lyman bullet mold. The mold produces hollow-based slugs so they fly true. Once you test pattern either shot or slugs, use the same load always. In the area of defense and law enforcement, for example, you want top performance and <u>reliability</u>. So once you find a load you like, it won't change. Same case, same powder, primer, etc. Think about this also if you have more than one shotgunner on your team: Full-length-resize everything; you need to be able to chamber each other's shells and you can't do that if you don't reduce the size of the brass bases.

DUPLEX LOADING

A major problem with shotguns involves pattern and range. It's nice to be able to let loose with a load of 6's and spread more than 300 pellets on the way to a target. But out of range, 6's will only sting. But when you load 00-Buck, the shot goes a lot farther---with feeling. Of course, you can only get nine of them in the shot charge (non-magnum) so you don't get a dense pattern spread. What's the answer? Load both. Any combination of shot will do. Some trap shooters load half

DUPLEX LOADING

00 Buck in front of 6's

One ball in front of 6's

Heavy ball behind 6's is a no-no.
It blows a hole in the pattern.

> When you duplex-load with two kinds of shot, remember to put the heavier shot forward. Otherwise, the lighter shot will slow down and the heavier will blast on through, which will put uneven gaps in your pattern.

8's and half 6's. That covers them for a bird that gets too far away before they get on it. Buck and ball loads do well for heavier game encountered at various distances.

Remember, however: If you're in the business of engaging, be it military or criminal enemies, don't engage a rifle with a shotgun unless you happen to be operating within your range, exclusively. If you engage a rifle shooter and he opens range on you (gets a 100 yard plus distance between you), prepare to meet the Lord.

As with rifles, good shotgunning starts at the loading bench. Choose good components, stay with the recommended load weights and primers and you'll be most effective.

III Chapter 1

MODERN RIFLE

Years ago, I won some gold medals in the Fort Ord Commanding General's Rifle Match. Because of that I shot competitively for the Fort, eventually made the Sixth Army Rifle Team and competed in a lot of rifle matches. At the more important matches some commanding general would give a speech. Several older shooters on my team would always mimic the general the day before; in a deep voice they would say: "It's not the tanks and big guns that win wars on the ground; it's the soldier with his steady hand, sharp eye and his well-zeroed rifle."

They can mimic all they want, but it's the truth. Later, in Special Forces, I knew it for sure. I used to think I could position myself with my rifle and some of my own special loads on the side of a hill—-and if they sent perhaps 25 soldiers after me from a thousand yards away, I would send 25 bodies to judgment. Since then, I've spent several evenings with the great Marine sniper, Carlos Hathcock. I don't care how many they would send after Carlos; he would bag them all.

The Corps credited Carlos with over 90 confirmed kills, but he knew the number was much higher than that. He shot a bolt action .308 with match grade ammunition. Together with his spotter, they wreaked so much havoc on the enemy in Vietnam that the North Vietnamese government offered a reward for him—-the equivalent of $10,000 U.S. dollars. Many people tried to earn that money; they all died poor. One wise old experienced mountain man came close, but failed when Carlos put a bullet through his scope while he was aiming—-at guess whom? Carlos. That only goes to prove an old maxim: Close only counts with hand grenades. If you want to be a hot shot, only the bullseye will suffice.

I learned a lot from Carlos, but I disagreed with one thing he did—-trusted his ammunition to shoot consistently. Apparently, all went well. But I would have carried my own powder and primers and put my own ammunition together. Lee loaders are just for that purpose. Since Carlos only shot one velocity and one bullet weight his rifle was limited even though he held it better and shot it so external ballistics were a shooting aid. I think he has a PhD. in wind reading and he stalks better than a cougar. Without a doubt, Carlos is the best sniper in the world.

Every year I hunt, I put together about twenty rounds of ammo I know will shoot on the dime. I drop the powder a little lighter than the weight I need, and then trickle powder, one granule at a time, into the powder scale dish to bring the weight to perfection. I am sure each cartridge shoots exactly the same. Carlos never did that. He had a match grade rifle. His action had been bedded into the stock and the weapon accurized so it shot groups tighter than one MOA. He spent hours at home in his living room working his bolt thousands of times with lapping compound so the bolt was super fast.

But what about the ammo? First, I am not too happy with the .308 for long distance shooting. Bullets with better coefficients are available. Also, when you load your own, you can mix and match the same bullet weights to obtain more or less penetration before expansion. If your target shows up in a military vehicle, why not use an armor piercing bullet? You don't get these at the five and dime. Tour the Army/Navy Surplus stores (where they sell my other books) to discover old rounds for .308 or 30-06 (M1-Garand). Pull the black tipped bullets. With Vihtavouri powder, build super velocity armor piercing rounds that would make a 6'-11" Ephesian armor bearer proud. Couldn't find one? Drill the nose of an FMJ round carefully so it accepts a diamond needle of an old phonograph. If the bullet stops, I think the needle should whistle right on through.

One night during dinner, I asked Carlos if he had ever missed. He answered, "More than once," and then told me about an incident that wasn't in his book. I feel certain about this: If Carlos had put together rounds with better ballistic coefficients rushing along at an increased 600 feet per second, the following incident would never have taken place:

He knew that a certain high-ranking North Vietnamese Colonel would come out just before dark to review his troops every night. So Carlos maneuvered for a day or so to get into position across the river from the colonel's Vietnamese camp. Sure enough, the troops were all waiting in formation. Carlos waited too. He was flat on the ground in grass near a gully he could crawl into and escape. If the Colonel showed up tonight, he would take the shot—a little over 600 yards over the river and into the target. Slight problem, however, the wind was whipping along at about 9 miles per hour from the side. With no range flag, and no shrubbery (remember, across the river) moving, it was hard to judge that wind. Finally, the Colonel showed, with his attaché in military formation—about six feet to the rear and six feet to the left. As the Colonel began his pep talk for the evening, Carlos breathed, let it out, relaxed, aimed through the scope by holding into the wind six minutes of angle, squeezed, and—bang! The bullet sizzled over the river and made hamburger—out of the Colonel's aide.

Of course, the Colonel had no sense of humor. The whole camp came alive. But bullets crossing water confuse people as to the source, so while Carlos was crawling quickly through his pre-planned escape route (a gully) the Vietnamese company peppered the trees close by and attacked in the wrong direction to kill the sniper.

If you want to own and control your AO (Area of Operations) absolutely, get a rifle and reload for it. A rifle is the king and ruler over shotguns and pistols—-unless your enemies happen to get within close range—-say, 75 yards. Open the range even a couple hundred yards and they may feel for you, but they never quite reach you. This will be your opportunity to feel like a preacher. The first dude to expose a part of his body will be made holey.

This is why I once taught my wife who drove a Ford Explorer to carry a rifle. She was gorgeous and the possibility of some drunken fool harassing her on the road was substantial. Cowboy country---Arizona, you know.

If attacked, the plan was: Drive off the highway onto a dirt road and lead them a merry, dusty chase for about several miles. After they would follow quite a ways, she would turn off the dirt road, go into four wheel, and put about 300-400 yards between them. Then she would stop, take cover, and disable their vehicle. Any further aggression? She was far enough away to shoot from relative safety and to pick off several aggressors.

That's why we clear cut the ground around Special Forces A camps. It discourages trespassers. I'm pretty sure that was the origin of the popular house sign now in use in this country: *"Do you believe in life after death? Trespass here and find out."*

On the subject of clear cutting, I can't tell you this story happened because we may or may not have been in a certain restricted place. So—-this is fiction.

As team engineer, I blew up whatever we needed to destroy and I built whatever we needed to construct. In this case we needed to clear fields of fire. So I was out with my chainsaw, clearing ground. In those days, officers assigned to Special Forces were not SF people; merely off-the-shelf Army ROTC or whatever. So, when I left a good sized tree standing, a 2nd Lt. told me to cut it again; "The order," he explained, "was to cut trees close to the ground." We argued. He didn't have a need to know and I didn't tell him why it had to stand that way—odd—-a three-foot-high stump about 250 yards from camp. I got some help from higher up and the tree stood.

A few nights later, the enemy attacked. I had aimed and sandbagged an M-14 with a special magazine. I'd pulled the black tipped AP bullets, added powder to the case and reset the bullets. During the battle, I merely squeezed the trigger. In the morning, I found holes clean through that tree. Surprised the lieutenant. It also surprised several bodies who had hidden behind it.

CHOOSING A RIFLE

Every day in America, men and women walk into gun stores to purchase a rifle. They hold the weapon up to their shoulders and take aim on a store wall. It has a new-fangled synthetic material stock—-light-weight durability. (Watch out in hot climates; synthetics can warp.) The weapon looks good, feels OK, the action seems smooth enough and the trigger clicks, so they buy it.

That's the wrong way to go about buying. Consider what this device is. It's a delivery system for projectiles. Doesn't it seem logical to consider first where you want the delivery? How fast? How far? How many? How hard? Night or day? Finally, how accurate (measured in MOA's) is the delivery ? That's a most important question.

So shopping for a rifle doesn't begin in the store, it begins in one of two books. If you'll buy store-bought ammunition, try *Gun Digest*. Consult a reloading manual. I've always been partial to *Hornady*. In fact, that's the book containing the information from which I build my trajectory nomographs, a trick I will teach you. *Hornady* manuals come in two volumes now. You need both. Volume II contains all the zero information you need for a specific bullet weight and muzzle velocity. Also---and this is critical for military---they list bullet energy. You can't get this information off a box of commercial cartridges; you <u>need</u> these volumes.

WHAT CARTRIDGE?

Having your rifle chambered for the same caliber as your pistol might appeal to you, but it isn't practical. Pistol cartridges are simply not powerful enough. Don't handicap yourself with a weak cartridge. For those who buy rifles for hunting, many experts have written dogma on what caliber is needed to take specific game cleanly. When considering animal size and caliber, ask this important question: Can it fight back? On the North American Continent large bears—-brown, grizz and kodiak—-are the only real threats.

The mistake most rifle buyers make concerning caliber is this: They look in the book and determine muzzle—-velocity and energy to get a basis of comparison. But cleanly taking game is not a function of internal or external ballistics; it's a function of terminal ballistics. To pick a personal game getter, you need to know what the energy level is when it strikes at the distance you intend to shoot.

If you hunt large bear (Alaskan browns or Grizz), you need a cartridge to deliver 2,500 FPE (Foot Pounds of Energy) at the target. For example, the .308 Winchester with a 180 grained bullet has a muzzle velocity of 2,620 fps and a muzzle energy of 2,743 ft. lbs. If you are shooting bear at the muzzle of your rifle, this cartridge looks powerful enough. But at 100 yards, the cartridge's energy picture changes severely. At that distance, the velocity has dropped to 2,393 fps and energy has decreased to 2,288 ft. lbs. Incidentally, it is neither normal nor smart to shoot bear at your rifle's muzzle.

Some hunters have reported shooting grizzly bear with small calibers, including the little .22 LR. Others have gone to try it; we're still waiting for their report. What should you do? Use enough gun. Take a good look at a cartridge ballistics chart. Find the cartridge you intend to use and check out its extended velocity and energy, out where your bullet strikes. My friend

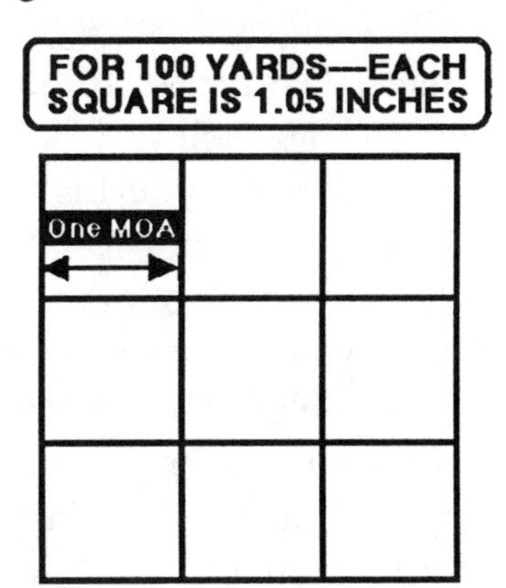

Micky Volk, an Alaska hunter, claims foot pounds energy is not the only requirement. He and many other experienced hunters believe caliber and bullet weight are needed to stop Grizzly. Recommended: Calibers larger than .308 and 3000 ft. lbs. energy out where the bullet will connect.

For large non-dangerous game, such as elk and moose, you should have a minimum of 2000 ft. lbs. energy at the point where your bullet strikes. Smaller game animals like deer require a minimum of 1000 ft. lbs.

MOA's AND THEIR RELATION TO DISTANCE

As you have read, MOA stands for minutes of angle. One minute will move the strike of the bullet 1.05 inches at 100 yards. So we express weapons' accuracy in MOA's. Personally, I don't want to own a weapon that can't shoot less than one minute of angle. Why? Because I may shoot as far out as 1,000 with it, where a one-minute-of-angle-error means the bullet could go ten and a half inches off target. That's not the worst of it either. Unless you can estimate range quite accurately, bullet drop and hold-over can be off as much as 6 inches. If you make a mistake when estimating range, or you forget to hold under correctly, your point of aim can be off as much as half a foot. When you add that to how your rifle might spray lead all over you'll shoot misses consistently. To the hunter that means go home hungry. To the soldier it means go home holey.

Where do you hunt or plan to survive? If that terrain is in the Northwest, then you may never get a long shot because much of the country there is filled with trees. Even so, you ought to buy a rifle that can perform at a distance. Will you be shooting through brush often? Busting through brush without serious deflection is a function of size; .30 or larger; weight; over 170 grains; shape; not pointed; and low velocity; down around 2,000 fps. Maybe you'll choose a larger caliber, perhaps a .358 Winchester. What will you be hunting and at what distance? Look at the tables. You need a bullet that will work for you on that target a certain distance far away.

But how much smack can your shoulder take? If you purchase a caliber you can't load down to a lighter bullet with a slower velocity, you won't want to practice. That's probably why not too many rounds are sold in the over .400 calibers. When you shoot them, it hurts. I once loaded a Weatherby .300 with a heavy bullet and made it fly as quick as the loading tables would let me. What happened was pure ballistic poetry in motion . . .

I made a mistake and shot it prone, and almost busted—- my collarbone.

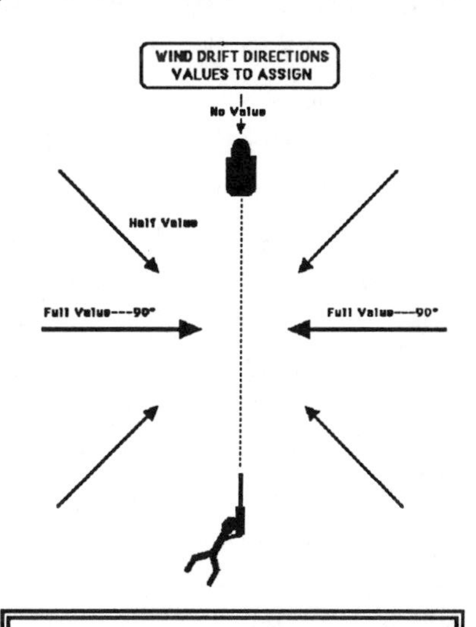

Note how the wind affects bullet flight. The lighter the bullet, the more effect the wind will have. Also, wind direction is critical. Turn your spotting scope into the wind until you see a boil; that's where it's coming from. No scope? Feel it, or look at some directional indicator, like a piece of green yarn you tie to a tree limb nearby.

To get speed, use a light bullet with a heavy powder charge. But such bullets drift in the wind.

How fast a bullet you need depends on how well you estimate range. A poor sense of depth perception can be improved; you can learn range estimation by using the 1-9 rule we taught in *EVERYBODY'S KNIFE BIBLE*. Then too, you can buy a range estimator, but that's just one more piece of equipment you have to carry and maintain. Assuming your range estimation skills are not so hot, you need a flat shooting weapon. You'll be looking in the book for something that shoots out of the muzzle around 3,000 fps. If you can't achieve that velocity and maintain it at a distance, your mid-range trajectory will be too high. Of course, some of us already own 30-06's or even magnum guns. Simply load them down; you can put 110 Gr bullets in a 30-06 and make them sizzle.

Besides the wind drift problem on light bullets, the little ones don't carry enough kinetic energy to the target. You need energy during terminal ballistics to cause the bullet to open up and perform. Thus, even though you reach out around 3,200 fps with a 22-250 or some other rocket, its lack of weight doesn't keep it going. So at 500 yards, you may land on target with insufficient influence.

Finally, all my expertise may be for nothing. Why? The decision might be made for you because you chose (wisely) a military rifle which shoots a caliber for which you can find scads of ammunition easily. Remember, if you shoot OP brass, you want your caliber to be popular. (OP's = Other People's—-seems to be a sought after brand.)

Assuming you chose the right cartridge for your needs, you begin to look for a rifle that will deliver it for you. Here you need to ask yet another set of questions before you're ready. How much barrel length? What kind of stock? Will you be sighting with iron or optic? What about magazine capacity? Finally, what kind of action? That's a critical question, because it has a lot to do with accuracy. If all your shooting will be done at short range, a lever action might be all you'll need.

DOES THE ASSAULT RIFLE REPEAT ERRORS?

In high school, we were all into hot rods. Guys would add carburetors, change cams and burn rubber. But there were other guys who couldn't figure out what to do with a wrench. These were the people who usually dressed up their engines with shiny stuff. Thus the expression, "If it don't go, chrome it."

Now we're older. We have all kinds of rifles and all kinds of shooters. The new expression is, "If you can't shoot straight, get a repeater." I thought about that when President Clinton pitched his ban on assault rifles. He said, "We can all count. Twenty, perhaps thirty rounds."

> My reflection: "Heck, if you **do** count you run around with some of *them Arkansas playboys* who can't shoot." Any decent shooter I know doesn't even give you a chance to count to one. All your friends are musicians.

Most of the people I run around with only need to squeeze a trigger once. Those kinds of people know for sure that if the round doesn't hit, it wasn't their fault; it was the weapon's. Therefore, you don't see them shooting pumps or lever actions. Until World War II, the lever action was America's favorite sporting rifle, perhaps because it's user-friendly for right or left-handers. I've owned several. When chambered for low-pressure 30-30 cartridges, such a rifle will last for life. I use my 30-30 as a trainer. Reloads for it cost only a little more than .22 rimfires, and you can always carry some righteous rounds—-just in case. Other lever actions with box magazines accept modern cartridges and a few offer a rotating locking lug to handle-high pressure cartridges. The lever action is not a bad choice, but you need more precision and longer range from your main squeeze.

Semi-autos chamber rounds for you as soon as you pull the trigger. Thus, you don't have to move your rifle off target to prepare for a second shot. Of course, before it loads a new round, it ejects the old brass casing. Where? Hither and yon. That's like leaving your calling card; anybody can look at your brass casings and know the kind of gun and ammo you shot. Also, you often lose your reloadable brass.

Someday, you will be faced against semi-auto weapons with only your rifle. Don't worry. It isn't how many bullets they shoot that makes the ultimate difference. It's how far you shoot accurately. That's why you reload for your rifle. When you control the range at which combat occurs, use a NECO-coated bullet with a good coefficient and aim your fire-lapped barrel with a steady hand, you're in control.

This is our rifle choice: The bolt action. Barrel length should be 20 to 26 inches. Shorter barrels are handier in tight areas such as vehicles and heavy brush, but a shorter sight base makes your rifle more prone to sight mis-alignment. Synthetic rifle stocks have a lot going for them, especially in the weight-saving category. Also, listen to the advice of Dan Norwood, gunsmith for a well-known rifle team competing every year in the nationals. Dan was my roommate in Dallas at the Shot Show and we spent hours together. He told me, "Composite stocks don't swell and shrink like wood ones. The actions are already bedded when you bolt them down. Therefore, you get more accuracy."

One fine example of a great bolt action is the Mauser Model 98, the most important rifle ever made. Most modern bolt action rifles trace their designs back to this strongly built, most accurate, military rifle. Over two hundred million Mauser-type rifles have been made and most sporting rifles made today are Mauser clones.

They're available in every caliber, barrel configuration and weight. The bolt employs a camming action and gives you leverage to chamber and eject sticky cartridges. For cleaning or inspection you can remove the bolt easily, exposing the action, magazine, chamber and barrel. Illuminate your bore from the muzzle end. Note this: Cleaning a rifle barrel from the breach is far better than the muzzle because you escape the problem of damaging the crown (at the muzzle) with the cleaning rod.

DAVE'S ALL-IMPORTANT SELF TEST

Test your practical field accuracy. Use 10-inch paper dinner plates as targets at various distances. Start close; place one each at 50 yards, 75, 100, 150, 200, 250 and 300 yards. Now try hitting these targets, five rounds each, from an offhand position. Later, paste over the holes or replace the plates to shoot again from a supported shooting position. Doing this tells you how good a marksman you are and teaches you to estimate range. Estimating range is a hard thing to learn, especially over uneven ground. With practice you can get better at hitting the plates.

Also, and this is critical, you'll determine your maximum effective range. When you factor in your rifle's inaccuracy, your shooting ability or lack of it, the bullet's external ballistic capabilities and wind drift, you'll know how successful you can be. You may come to the conclusion that a 300 yard shot is too far, or maybe 200 yards exceeds your capabilities. To improve, do three things: Upgrade your rifle, hand load an improved cartridge, and tighten up your own shooting ability. Of course, practice shooting.

Use our MOA grid to improve position and determine wobble. If in sitting position, for example, you're MOA wobble is excessive from side to side, <u>make sure</u> your elbow is directly beneath the weapon. Do you wobble high and low? Work on breathing control. Perhaps you're supporting too much rifle weight with your non-trigger arm. Dry fire enough to get your trigger finger a PhD. Load your own ammo and learn your capability.

If you hunt in open plains you may choose a magnum cartridge because it will give you enough energy and reasonable trajectory at 300 and 400 yard range. The draw back is this: They operate in the 60,000 to 65,000 psi pressure level range which is hard on shooters, barrels, and brass. Brass really starts flowing at 70,000 psi and this causes your brass case to stretch. Case life is reduced as the case thins and you have to trim away case length. Because of magnums' high-pressure levels, the powder gases burn out barrel throats, causing an accuracy loss after about 3000 rounds. By the time you've put 10,000 rounds through your magnum rifle, inaccuracy will be noticeable.

To avoid these problems, buy the magnum rifle and download for it. You don't have to shoot those high-pressure loads all the time. Any good rifle should yield practical accuracy up to 20,000 rounds if you shoot molded lead bullets at around 1,900 fps. So supplement your ammunition with a supply of molded lead bullets to get long life out of your rifle. To make the lead harder, anneal it by adding tin and dropping the hot round into cold water. The ability to mold your own bullets may someday save you. By reducing your velocity and bullet weight to a lighter load, you can shoot your magnum rifle **ammo forever**. Of course, there may come a day when you need to reach out and touch something long distance. You only need to practice shooting Alaskan bear whumpers enough to know how they perform. Then, once you learn, build your trajectory nomograph for these rounds and carry a few in your ammo wallet. We designed the load/performance forms so you'll have a permanent record. Should the need arise, you'll be ready.

No matter which rifle you purchase, it doesn't come ready to shoot out of the box. It needs improvement. Just for starters, all weapons made in the U.S. are ostensibly for hunters. That means the sighting systems are for daytime shooting only. What if you have to use the weapon for defense? I suppose you can ask criminals and gangs to operate during daylight hours only. Of course, there's lots more. How do you carry it? Should you change the stock? Does it need accurizing?

In the next chapter, we tell you how to improve this rifle so it becomes something special. When you finish improving your rifle, it will shoot farther, flatter and right on target. Furthermore, you'll know how it performs as if it were part of you. That's next . . .

When your scope fogs or gets knocked out of zero, you'll need back up sights. The peep or aperture sight on the back of this mount is the most accurate.

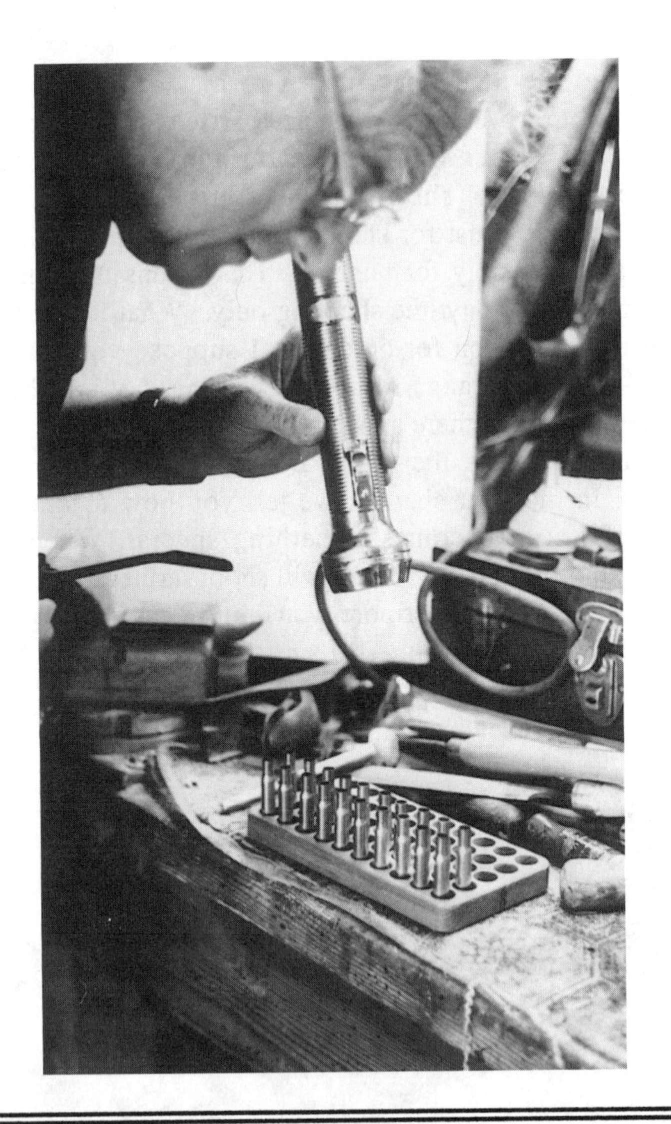

Check twice--Live once. With a flashlight look down into the mouth of each case to insure all powder levels are exactly the same. Also, this procedure insures that you don't seat a bullet in a primer-only round.

III Chapter 2

RIFLE ACCESSORIES & IMPROVEMENTS

ACCESSORIES

Because the rifle is a long-distance shooter, it's the weapon of choice for people who go into the field for a period of time. It's also the weapon to have in a lengthy combat struggle because it will out-range a handgun or shotgun. To feed your family it's the ultimate game getter because it can reach out hundreds of yards to pre-condition meat for your freezer. That's why military and police snipers use a rifle.

When I was shooting for the Sixth Army rifle team, I used to show up at the range looking like a second-hand junk man. In the field, I carry even more because I stay out so much longer. To be a good rifleman, you need to bring along other devices and goodies. How, for example, can you examine a target or read the wind without a spotting scope? Also, since the scope is normally a 20 X, you'll need a tripod on which to rest it.

Begin with a rucksack. Get one to fit your physical ability and body size. Mine is an exterior frame Jansport, and the company has repaired it without charge. I like the exterior frame because I operate most of the time in hot humid climates, and I want air circulation between my back and pack. Make yours bullet proof if you're tactical. Buy Kevlar panels to stop incoming rounds from penetrating the sides of your pack. With your pack on the ground and you behind it, you'll be relatively safe.

Back Pack as Firing Port

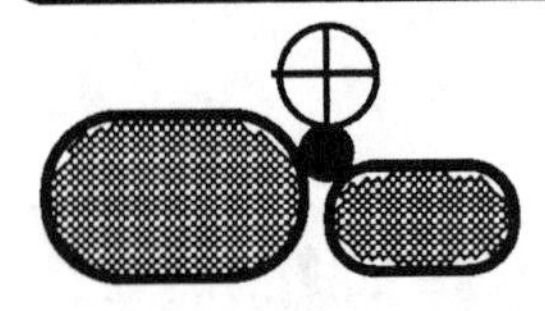

The way things ought to be. Nobody makes a rucksack this way, but fix yours to rest a weapon like this and your long distance shooting will be joyful.

Choose a backpack model with top and bottom compartments separate. Thus, you can lay your rucksack on the ground and place your rifle between compartments so you can use it on the ground like a shooting port. Sew bullet loops on the outside of your ruck, points down. Label the compartments each to hold a round for a particular purpose—-armor piercing, BPBT, (Banana Peeler, Boat Tail) soft point, round nose etc.

In an outside compartment, store enough ammo to save General Custer. Also, consider carrying a nine inch long piece of old pant leg, sewn up on each end to contain either sand or kitty litter. The liter is light weight, but you can use it as a sandbag to steady your rifle. If you'll be crawling through the woods on your belly like a reptile, pack some soft rubber elbow and knee pads. Animals are wise to the sound of plastic pads knocking against wood and rocks.

Pack a kit to keep your rifle clean and lubed. You need two compasses (see *GREEN BERET'S COMPASS COURSE*). You can drill your stock and press-fit a back-up compass right into the hole with glue. If you use the PAUL system you don't need a map other than for terrain analysis.

THE CRITICAL DAYPACK

A small daypack fits inside your major ruck. Note: This is not a small canvas carry-bag with a handle. It has shoulder straps because you need both hands free anytime you carry a shoulder-fired weapon. For a climb or a long stalk, cache your big pack and use the small one. Mark the cache spot well with a bright cloth somewhat away from the pack so someone else doesn't find it. From the bright cloth to the actual cache spot, you'll have to memorize directions by travel vector.

In the daypack, carry optics, parachute cord, signal mirror, compass, game bag, folding (Blackjack) knife, space blanket, poncho liner, perhaps a small pillow and a small flashlight with extra batteries. Pack a few camouflage or olive drab bandannas. One camouflages my neck and I sterilize a couple in case I need bandages. Include plastic restraints* for making a variety of outdoor furniture, shelter and goodies. Feed sacks cost about twenty five cents each, and will furnish you with a wonderful above-the-ground bed. Sleeping on the ground is a no-no. (See *EVERYBODY'S KNIFE BIBLE)*.

*Plastic restraints are particularly useful for making a tripod. Tripods are more stable than bipods.

You don't have to carry them on your rifle, either. You can make one in any area in which trees grow. With a sharp and hefty knife such as a Blackjack Mamba (*EVERYBODY'S KNIFE BIBLE*) you can cut a tree limb so the ends are somewhat square. You need three pieces, each long enough to come up to shoulder level when you shoot sitting, offhand or prone. Two-thirds of the way up from the ground, wrap these with a plastic tie; then spread the pointed ends out to make a tripod and jamb the whole thing into the ground. The pants leg filled with kitty litter sits on top of that and supports your rifle so it doesn't move. You'll be amazed at the lack of wobble.

HAND, ELBOW AND KNEE COVERINGS

You need gloves. Aviator's gloves are excellent, but several others will do fine. Cotton gloves, for example, are the choice for most hunters. I also prefer to use a glove insert, especially on my left hand because it keeps the upper sling swivel from pinching my hand when I use a loop sling. Cut off the trigger finger of your glove; with velcro, it's re-attachable. Besides gloves, cut out a rubber mat and fix it up so it will attach with velcro to the outside of your gloves at the palm. That way, you can grab any rock or rough surface easily. With knee and elbow pads, you can train yourself to go over rough terrain on hands and knees with some decent speed while dragging your rifle behind you. Think of the possibilities: In 125° desert heat, you could approach and stalk anything by crawling quickly up a rocky gully. The same trick would work in a dry rock wash during the summer in a jungle. Thorny vines couldn't bother you.

CARRYING CASES

Cradling your rifle in your arms at the elbows is the worst. Elbow pads are OK, but your rifle catches on every vine and bush God ever planted. Make a drag bag to haul your rifle through the woods with the muzzle pointing behind you. (Handloaders sometimes experience auto-detonation.) You want it water-proof but you also want it to shed moisture, so GoreTex is a good idea. Make it somewhat sturdy and well padded inside. In particular, don't let the scope take punishment; otherwise you could destroy your zero. On a difficult stalk requiring you to crawl over much terrain, a six-foot towing strap with a shoulder loop lets you drag the rifle easily. Any extra padding you can place in your drag bag around the scope will help. Also, watch the sound. If the bag is cordura and you drag it over rocks, you alert the whole animal kingdom to the presence of something foreign. You can use your handgun to take care of short-range surprises.

ON THE JOB REFRESHMENTS

Get dehydrated and your eyesight won't be too sharp. Carry a canteen and a water filter. I don't go anywhere without my water purifying filter in my rucksack. With all the infestation in mountain water these days, you'd be safer drinking out of your toilet.

> Carlos Hathcock took C-ration cans of cheese and peanut butter with him because he didn't want to defecate during a three-day stalk. The man is amazing. On a hunt, he gets into a psychological bubble. He eats small amounts, sips water from his canteen and moves in slooow, never-ending motion.

Freeze dried foods you can mix with filtered water are practical. Food high in protein such as Chia Seed, Peanuts, etc. is a good idea. Eating the wrong foods while hunting can be like getting stuck in freeway traffic after eating bran muffins and drinking coffee.

COLD WEATHER GEAR

The temperature drops severely in many areas.

Check yours. Even on a long-range solo mission, you can carry a short wave radio with earphones to get a weather report. You may have to climb into some higher elevations, so carry some poly underwear and a sweater to make sure your upper torso doesn't chill. Also, remember that your head sheds 17% of your body heat. A black knit stocking cap under a floppy jungle hat keeps heat in your body.

PLAN TO STAY OUT OVERNIGHT

Since big game and enemy troops come out at dusk, always plan to stay out overnight. Much of what hunters bring home is bagged just after sunset. By the time you field dress it, darkness is falling. To make it back to car or camp safely before dark, you will have to bee-line; you don't have time to back-track. Either use a map or lay out a P.A.U.L. diagram.

Even so, you may have to stay out overnight. Rolled underneath your daypack, add some extra warm clothing, some rain gear, water and food. More important than food is good sleep. Bring a lightweight, but warm sleeping bag. For me, a self inflatable pillow helps a lot. You absolutely need rest. Tired hunters don't see well, make mistakes and sometimes suffer injury from accidents.

Keep repellent in your daypack. Otherwise, you may stay up on an ambush and swat mosquitoes all night long. Sew pieces of mosquito netting on the outside of facial spandoflage to help with insect management. Drink a little cider vinegar with your water too; bugs hate that kind of sweat. If you carry screw-in tree steps, your daypack would be the place.

BODY WASTE FUNCTIONS WITHOUT MOVING

It's a good idea to bring a make-shift potty with you if you don't want to get up in the cold during the night or if you are on a mission which requires minimal exposure and slow movement for long periods. Stump-sitter type stalks or law enforcement airfield surveillance are two such instances. While lying on the ground, roll to one side and drain into a plastic bottle with a screw-tight cap (GatorAide or Orange Juice). Dig out a little cat hole for the contents; cover it. One health note: Don't share the bottle. Even two healthy people carry bladder germs. One or both of you won't be able to equilibrate the other's bacteria. Result: Bladder infections all around. Even if you have the right antibiotics with you, your mission may be compromised.

Whether on a long stalk or approaching a military target, this is not the time to have to respond to emergency bowel demands. Learn to defecate while in the supine (on the ground) position. You dig a hole and roll over it so you face the clouds. Roll to one side and lift one knee for clean up. Then replace the dirt so nobody with a sharp nose is drawn to your tracks.

OPTICS

Optical overlap is important because it allows you to detect targets over a much larger area of ground. High-power optics enlarge targets for close study. Low power optics cover wider fields of vision. You really do need a spotting scope for long-range shooting because it tells you what the wind will do to your bullet. (See *EVERYBODY'S OUTDOOR SURVIVAL GUIDE)* Carry a small tripod if you have the room. Otherwise use a kitty litter bean bag you made on top of a tripod similar to the one you made for shooting. You can also hang the scope off a tree limb with parachute cord, then pull it down firmly during use.

When you hunt in pairs, overlap your optics. One person uses wide angles, low power. The other—-high power binoculars. For the hunter, 7 X 35 wide angles will let you look for movement in a large area. Mini-binoculars are becoming popular; I think I own three pair, one of which is 10 X. Zoom binoculars are also nice. But higher power optics always shake as if you had the St. Vitus dance. Steady them down by using a rest so you can see things more clearly.

CLOTHING

Sew together a camouflaged outfit you wear in the woods. Turn some GI camouflage BDU's or Tiger Stripes inside out and modify the pockets. Sew rubber padding in at the knees and elbows. Cut out the armpits and crotch, then sew in netting to keep your body from overheating. Cover the whole suit with ASAT (All Season, All Terrain) camo cloth or perhaps cut up one of our hammocks and sew it onto the main suit. From the hammock material, hang camo cloth in strips or gunny sack material dyed in three woods colors. Bring a spandoflage face net. When you wear this suit and cover other exposed areas, such as the backs of your hands and your rifle, you'll be undetectable---until you move. Camo paste is important because it covers and conceals all the areas left exposed after you dress up. One final word of advice: don't forget fire retardant. You'll probably be dressed up in all this hanging gunny sack material and ASAT cloth someday near an open fire. Want to wash your hunting clothes? Use special soap that removes ultra-violet light emissions. Otherwise, a deer sees you---pretty much the way you see a road sign in the dark.

KEEPING UP WITH THE WORLD

A short wave radio is a must for a long-range mission. The world changes. I tutored an American helicopter pilot, Ben Springer, who flew in West Africa. His short wave radio warned him not to land near Monrovia after a revolt next door in Sierra Leone. Short wave radios are lightweight, compact and easy to use.

They pick up broadcasts from everywhere. Most of them contain weather bands so you can get the latest report. That's important information if you're on your way to penetrate an enemy line. You want no surprises.

NIGHT ACCESSORIES

Buck now sells a light that puts out super power with minimum weight. I also like the mini-GI flashlights powered by two AA batteries. If you are tactical but watching out for animals in the night, bring two half dead batteries. You don't have to power up your flashlight to illuminate any animal, the slightest light will make reflectors out of his eyes. In fact, if you can see the whole animal clearly, you're using too much light. When you flip the light on, have your fingers (closed and extended) over the lens. Open your fingers slowly to let only enough illumination required-—no more! An example would be a jungle operation where you don't want to be discovered. Light and sound discipline are just as important as sight (camouflage) discipline in the woods. Not only will they never see you; they will never hear you.

> Once upon a time, some yo-yo shot his rifle with an obstruction in the barrel. That barrel was made of hard steel, so it blew up. I forgive the yo-yo. But I can't forgive his attorney, who argued that hard steels in barrels are dangerous and that the maker should have known that soft steels would not have caused his poor client such injury. So now, rifle barrels consist of soft steel.

RIFLE IMPROVEMENT

New, out of the box, a rifle represents about half of what it can be. You supply the other half. You need a rifle that fits you perfectly, has a smooth and light trigger that will discharge with the slightest of pressure, and shoots exactly where you aim it. New ones don't always do that because the inside of the rifle barrel, the bore, isn't straight and smooth. Minor bore bumps constrict the bullet so hot gas can escape around the sides and a little jacket material is roughed up on the bullet so it wobbles.

THE ALL-IMPORTANT BARREL

When manufacturers cut in the rifling, the land's and groove's surfaces are rough. If you were in a microscopic Disneyland cart riding down the rifle bore, the cart would bump up and down in places. Those places often squeeze a bullet on its way out. After having been squeezed a bit smaller, the bullet bounces back and forth in the rest of the barrel as it speeds away toward the muzzle. Hot gasses pushing the bullet down toward the muzzle leak around the bullet edges. Just as the bullet clears the barrel, some high-pressure hot gas explodes past it, nudging it from the rear so it flies out a bit crooked. In addition, small pieces of the bullet scrape off in rough places. After leaving the muzzle, those places will catch in the wind and cause bullet yaw.

The <u>heart</u> of rifle accuracy is in the barrel. If only there were some way you could make the inside of your barrel mirror smooth and remove the tight spots, then none of the above would happen. <u>All</u> of the hot exploding gas would stay behind the bullet, and push it evenly out the muzzle. Happy days---every round you shot would fly true.

No barrel new in the box will shoot like that, but I know how to fix it. What if you took a soft lead plug, imbedded a fine rubbing compound on the edges, and pushed it through the bore? It would polish the lands and groves. As you pushed it through, you could feel the tight spots. The polishing grit on the sides of your soft plug would gradually sand the rough spots down as if a tiny tidy-bore man were inside with sandpaper. Once everything was mirror smooth and all the rough edges and bumps were gone, your barrel would shoot like a woods king. Could we do it? Nope. Hand lapping requires expertise. You have to use the right grit on the lead plugs, and you don't dare touch the lands at the muzzle with the rod. A high percentage of barrels have been damaged during this process.

BEST: FIRE LAPPING YOUR BARREL

What's the real answer? Fire lap your bore and polish bullets. Write to NECO 1316-67th St., Emeryville, CA 94608. 510-450-0420. For $2, they will send you a PIG, Product Information Guide which teaches ways to make your barrel a precision bullet-placer. They sell a patented kit containing bullets of various grits and a moly-coat kit to make bullets stabilize earlier and slide through the air much better (improved bullet coefficient). When you fire these (slow-velocity) soft lead rounds through your barrel, magic happens. The inside of your bore will become a smooth, mirror-like surface. No gas flies by the bullet and nothing scrapes bits and pieces of the bullet jacket. Thus, every bullet hits the mark.

CHECKING THE CROWN

At the end of the bore (muzzle) it is crucial that the lands be in perfect shape. Also, each one of them must end at exactly the same distance from the bolt face; otherwise, one of them gives a little gotcha to the side of the bullet just as it leaves the muzzle. That de-stabilizes the bullet and makes it wobble as it flies forward. Thus, every time you fire, you get a new and surprising bullet hole---sometimes not even near the bullseye.

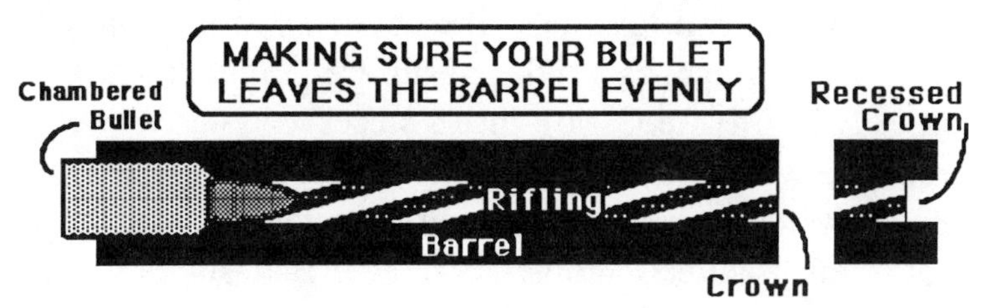

With a high-power magnifying glass, check your crown. Look for abrasion (caused by a steel cleaning rod, excessive wear, or anything not uniform). Each land must be exactly the same. If you suspect anything wrong, a gunsmith may go to the extreme of cutting a small piece off your barrel and recrowning the muzzle.

If you decide to recrown your muzzle, you might as well recess it. That keeps it perfect if you happen to drop your rifle and it lands muzzle down on some rocks. Also, once you have inspected your crown or had it resurfaced, either use a guide to keep your cleaning rods off it or draw your cleaning brushes and patches from the chamber (bolt removed).

STOCKS

The wood stock on many commercial rifles causes trouble. Why? It absorbs moisture. When that happens, the wood swells and puts pressure on your barrel. That doesn't mean it bends the barrel, but it does mean the barrel is restrained from "whumping" when you shoot. Result: A new effect on every bullet you send to tour the countryside; no two of them visit the same place twice.

When you replace the wood and buy a composite stock, you reduce the weight, accurize the weapon somewhat and customize the stock to fit you perfectly. Cut the stock to fit only you. You want a proper length of pull, measured from the trigger to the finished butt; also, consider the length of your neck. If your head sits tightly down on your shoulders, a shorter length of pull is in order. A good gunsmith will help you. Bed the action so the barrel is full floating. It's better if your composite stock has a rough, camouflaged surface. Some come with adjustable (Monte Carlo) cheek pieces you can adjust so your eye will always be in line with your sights or scope.

With the best ammunition, an accurized rifle, and a hot, well-practiced shooter, a trigger that slides toward

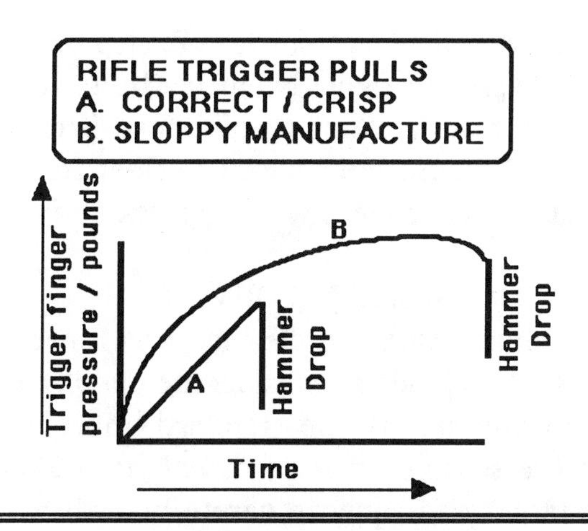

TRIGGER PULL DIAGRAM

On a bad trigger, (B) you keep squeezing and you feel metal-to-metal scratch. Then the pressure reduces somewhat just before the hammer drops. Yuk! A well-made trigger (A) is smooth and doesn't require as much pressure. You simply increase the squeeze and the hammer drops crisply by surprise. Close your eyes and squeeze; you'll feel it.

release with a scratchy, heavy pull will cause a shooter to punch a hole in nothing more than air. Bad triggers make you destroy sight alignment just before the bullet leaves the barrel. Late model triggers are pretty fine on all the major makes. Some old rifles have triggers like bumpy ski lifts, however, and require gunsmithing. Check the price of the work. Sometimes you can buy an superb after-market trigger to replace the scratchy one giving you trouble.

BEDDING YOUR ACTION TO YOUR STOCK

If the action jiggles around in the stock bed during shooting, bullets will strike in some amazing places. Brownells offers "Glasbed" compound to obtain a uniform bond between rifle and stock. The NRA has published articles that will take you from A to Z, step by step so you get the whole process right. No matter from what material your stock is mad, creating a solid bond between action and stock will cause your barrel to float free and your action to stand as one piece with the stock during and after recoil.

SIGHTING DEVICES

To hit anything at a distance, you'll need accurate rifle sights. Two kinds to consider are optic and metallic. Scopes, of course, are the standard improvement for sighting. The scope shooter needs only to place the cross hairs on target and squeeze carefully. Also, since the scope gathers light to make dawn or dusk shooting possible and target identification easier, it's essential for the survivalist rifle. But scopes aren't perfect. They're fragile, easily knocked out of zero and they can fog over in some weather conditions. That's why iron sights installed <u>and zeroed</u> on your scoped rifle are good for back-ups.

Scopes also can detract from good shooting ability in two ways. Too much magnification can give you a false impression of how far away your target really is. High power seems to magnify your wobble. When your crosshairs jump around on the target, you have a tendency to fire the rifle as the crosshairs pass the target. This will frequently cause you to jerk the trigger and destroy the sight alignment just as you fire.

To keep scope and iron sight mounts secure, use Locktite, fingernail polish or varnish on the screw threads. The best is low strength Loctite, for you may want to remove your mounts, change eye relief, etc.

Besides buying a good quality scope, you must select three important items, <u>magnification, reticule, and mounts</u>. Depending on where you hunt, the best magnification is between 2.5 and 6 power. Variable powered scopes are best, with powers ranging from 2-7 or 3-9. The lower the power, the wider the field of view and the more light gathering capacity. Also, low power enables you to track and shoot a fast moving target. If you need to inspect something closely, you can turn the power up. A 40mm. or larger front lens also will give you more light gathering quality. Of course, those bigger scopes require higher mounting.

What's the best reticule? Most popular is the four-plex. The post reticule shows up best in the dark woods. Still, the disadvantage with the post reticule will be obvious when you have to take a long shot and hold over because the post blocks out your target. Hunting in poor light with plain crosshairs or dots doesn't work well because those reticules fade in dark backgrounds.

The most ideal setup for a scope mount is one for which zero is the same after removal and remounting. Weaver and Warne both make those. Some scope mounts (Warne) also accept an aperture (peep sight) you can fasten to the mounts after you remove your scope. Good idea. Most other brands cause you to lose your zero completely after removal and reattachment to your rifle. Incidentally, if your scope becomes damaged during transport you'll have to rezero your rifle.

Eye relief refers to the distance between the eye and the scope. Too far away? You only see a little hole of light in the scope. Too close? Rifle recoil is going to cause the scope to smack you in the eye. Most shooters have their scopes adjusted for proper eye relief for the offhand position. If you could shave their right eyebrows, you would find a half moon scar over their sighting eye. Dave's eyebrows are scar free, but Don has one of those scars and guarantees the pain to be unique. You can get one too. Just set your eye relief while standing, then sit down and shoot with your eye much closer to the scope.

Solution: Set eye relief for sitting position, and gooseneck it if you have to shoot while standing.

Many good high-power rifles are sold without metallic, or iron sights because the scope sight is so popular. For the survivalist or hunter who's scope is damaged or made useless by weather conditions, iron sights are essential. You can get two types of metallic rear sights—-aperture and open. Normally the more precise peep sight attaches to the rear of the rifle's receiver. These have minute-of-angle adjustments for both elevation and windage.

Open sights are normally mounted on the rifle's barrel just ahead of the receiver and provide the fastest target acquisition for snap shooting. They need not be removed if a scope is also mounted on the rifle. The peep sight is reliable and accurate enough for 500-yard targets. Just make sure your sights are perfectly aligned and squeeze—-straight to the rear gently. If you lose your sight picture while squeezing, don't let off on the trigger. Just hold the pressure you have, get your sight picture back and complete the squeeze until the weapon surprises you by firing. You must educate your trigger finger. It takes practice...

SLINGS

Equip your rifle with a rifle sling of either nylon or leather. When you need both hands, you'll find a rifle sling invaluable. More than that though, I wouldn't use a rifle without a sling because it helps you shoot tighter groups and therefore adds accurate range to the weapon. The whole idea in rifle shooting is to tighten your MOA bullet placement. So you load special ammunition. You accurize your weapon. Zero it for long-distance and memorize the short-range nomographic mid-range trajectory bullet rise figures. Then—-improve your ability; learn to hold perfectly still with reduced wobble.

How? By using your sling and measuring your wobble on our grid.

By moving the sling frogs, you fashion a military leather sling into a loop which you circle around your arm just above the bicep. With your left (gloved) hand jammed in tight against the upper sling swivel, the weapon becomes almost motionless. Both leather and nylon slings can be used as hasty slings, which also will prevent your weapon from wobbling. Using the sling wrapped around your arm in hasty fashion enables you to balance the whole weapon in your arms is if cradling a baby. Make sure to lift on the pistol grip. Practice this and get used to it by using your deltoid muscle; you hold the weapon in your shoulder without supporting any substantial weight at the forearm. After practice, while in the standing position, you move left and right or up and down by moving your whole torso. If you're wobbling all over the map from the standing position, probably you are supporting too much rifle weight at the forearm.

Think about a second sling---a shoulder strap. If yours is an assault rifle, you want the shoulder strap to allow the weapon to hang down with the barrel level. As with shotguns, you accomplish this by putting matching velcro halves on the sling and on top of the shoulder of your shirt or jacket. Use a carpenter's level on the barrel before you set the velcro. With the velcro engaged, your weapon will always be level. Thus you eliminate high and low misses. With an automatic weapon, you use tension on the strap to keep the weapon from climbing. The net result with semi-automatic firing is that you send level rounds across the terrain from left to right about 300 yards ahead of you. Several shooters in the same level and slung mode should keep the streets clear. Therefore, riot situations (such as they have in Israel), would be stopped cold by horizontally firing rubber bullets. The velcro on the sling and shoulder could be adjusted much the same as a traverse rod on an auto weapon.

CLEANING

No matter what, you have to keep your rifle clean. Failure to do that will leave residue in the bore to be hit and smashed to the sides by a fast moving bullet. Some rifle shooters have this motto: Nothing more to clean the bore than Hoppe's. That's not so bad when you consider all the things you can destroy by over-cleaning or bad treatment during the process. An unlubricated bore brush will put scratches in the bore. If you rub a steel (Army surplus) cleaning rod against the lands of the bore at the muzzle, you will send some bullets to new zip codes. Teflon-based oils in the barrel cause inaccuracy, too.

***

Why settle for a new out-of-the-box rifle which can only shoot half as well as it could with your tender loving care? Add the accessories you need to be a long distance shooter. Acquire the other goodies to help you live in the field for long periods of time. Spend some time and effort on your rifle to improve the stock, trigger and bore. Once you do these things, you and your rifle can control a lot of land.

III Chapter 3

RIFLE---SHARPSHOOTING AND TACTICAL USE

I wonder if the phone company knew they were promoting rifle marksmanship when they came up with the popular slogan: "Reach out and touch somebody." That's what a rifle does. AT&T should be against gun control because rifles prove beyond a shadow of a doubt, "Long distance is the next best thing to being there."

I've won trophies and medals in thousand-yard shooting matches. You can too. Once you can hit from a thousand yards away, you own all the food on the farm. You also can enjoy relative safety on a military mission in a foreign country.

Notice the underlined word, hit. That's the opposite of miss, and to make your rifle hit always you need to eliminate all the causes of misses. What goes wrong when you shoot and miss?

First, you can experience interior ballistic difficulty—-you didn't load the round properly. Different powder changes cause bullets to fly high or low. If you don't adjust your bullet seating die to position the bullet precisely for your particular rifle, too much free travel before the bullet catches in the lands of the rifling can make it shoot erratically. Perhaps you let the round heat up in a hot chamber, so the bullet push is too severe, which makes the delivery like a high pitch in baseball. The rifle itself might not shoot well. If the bolt face is machined at an angle; if the barrel and action flop around in the stock; if you don't fire-lap your barrel to keep the bullet from rattling down the bore; or if the crown at the muzzle is sloppy it causes one land to stay on the bullet longer than the others, your rifle won't shoot consistently.

Second, the exterior ballistic phase of delivery may not go well. Just as you shoot, hurricane Myrtle decides to blow across the bullet path. The sun goes behind a cloud on a long shot which slightly changes your sight picture. You misjudged the range and failed to compensate with enough hold-over elevation (not the way to go anyway) so the round lands in the dirt before it reaches the target.

Finally, you may have developed some bad habits when you learned to shoot. Just as pilot error crashes most airplanes, shooter error crashes most bullets. Poor marksmanship makes bullets fly all over like a bad dancer trying to cha-cha. Any position which requires excess muscle tension will increase wobble. Rifle recoil is painful to a shoulder with little muscle, so some shooters flinch.

Suppose you reload rounds which can out-shoot the Lone Ranger's silver bullets. Suppose you modify your weapon and accurize it until it reaches out with enough feeling to make a psychologist proud. Even with all of this, you have to BRASS. That means: **B**reathe properly, **R**elax like a sleeping siesta bum on valium, **A**im with sight alignment to make a space engineer envious and get a sight picture Michelangelo would love to paint, **S**queeze with a PhD. trigger finger, and **S**urprise yourself when the primer pops.

When you can shoot like that with a space-cadet cartridge you built and it flies out of a flawless rifle, then camouflage yourself like a puffer fish in a coral reef and move like a bushmaster snake in the Panama jungle, you own the woods. They'll never see you, hear you, find you, and any enemy within a thousand yards better pay his life insurance premium. Both your ammo locker and your meat freezer will be full. It's the essence of ***GREAT LIVIN' IN GRUBBY TIMES.***

LEARNING TO SHOOT FROM POSITION

If you've never shot anything before in your life, get a coach; take lessons. Basic shooting ability is not something you can learn easily from a book. Although our wobble grid is a big improvement in teaching tools, somebody should coach you to see if you jerk the trigger or breathe while shooting. Your coach needs to adjust various shooting positions so you support the weapon to eliminate wobble. Along with many other fine points of shooting, perhaps he will make sure you don't lose your position when you eject and load.

Once past all this, however, you have some things to consider. Even before we get to BRASS, you need to learn about offhand, sitting, kneeling and prone, and the basic **wobble** each new position causes. Wobble is the word we use in this book to designate your natural inclination **not to hold still.** Your body is alive. Your heart beats. Your muscles contract. Your toes press down and release constantly to keep you in balance. All this and more, causes the rifle to move so that the aligned sights appear to sway back and forth and up and down across the bullseye.

If you ever shopped for a home with a boring real estate salesperson, you heard expressions such as, "this is the living room; this is the bedroom," etc. I get the same feeling when I read information about rifle shooting when the writers indicate, "this is offhand; this is kneeling," etc. They then explain such goodies as, "make sure your elbow is directly beneath the weapon, and, squeeze the trigger carefully, straight to the rear."

That's all good advice, but one important thing they don't explain is this: You can measure any shooter's wobble. You measure your own by using our wobble grid sheet. Simply sight on the bullseye and remember the outer limits to which your point of aim meandered. Add left and right deviation. Add up and down deviation. When you're finished, you should know exactly how much wobble you generate from each position you use. You should also be able to tighten that wobble by using measured wobble to make improvements to each position. With a coach, attach a laser sight to your rifle. From 25 yards away, shine the laser on transparent graph paper with squares .2625 inches each. The coach watches carefully and records how much wobble a shooter produces. The wobble graph is highly important because it relates wobble to cause. You can reduce wobble when you tighten your sling, change your breathing, adjust position, etc.

The human body wobbles more or less while in different rifle positions. **It is critical to know this.** Remember, MOA means Minutes of Angle, and one minute equals 1.05 inches at one hundred yards, 2.10 inches at 200 yards, etc. Although it varies from shooter to shooter, we know approximately how much each position produces. It goes like this:

Offhand	5 to 10 MOA
Kneeling	4 to 8 MOA
Sitting	3 to 7 MOA
Prone	2 to 6 MOA

WHAT GRIPES ME MOST ABOUT MILITARY INSTRUCTION!

Even today, small arms instructors stand up in front of recruits and say, "This is your rifle. **It** has an effective range of. . ." That's bull! Rifles don't have an effective range; shooters, cartridges and improved barrels do. You can make your rifle shoot farther if you tighten it up, and you can use ammo that will do much better at longer ranges. To be fair, it's not the instructors' fault when they tell you that; the military civilians (who often are WIMP's) publish that dribble.

The military needs to learn: Weapons are as effective as you make them, and each person can make his own a super-shooter. You want to be effective? Learn to shoot; build super bullets; improve your rifle. You really can out-shoot the world if you do all three!

HOW FAR CAN YOU AND YOUR RIFLE SHOOT?

Suppose you see a target and you decide to shoot from a standing position. Your wobble is average; say six MOA. That means the bullet can strike as much as 6.6" (inches) away from your intended point of aim for every 100 yards of distance you shoot from the target. How about bullet deviation? Your store-bought ammunition is only one MOA accurate. Suppose your rifle shoots only 1 (one) minute of error; that's commonly thought to be accurate enough. Note this: every MOA error from a

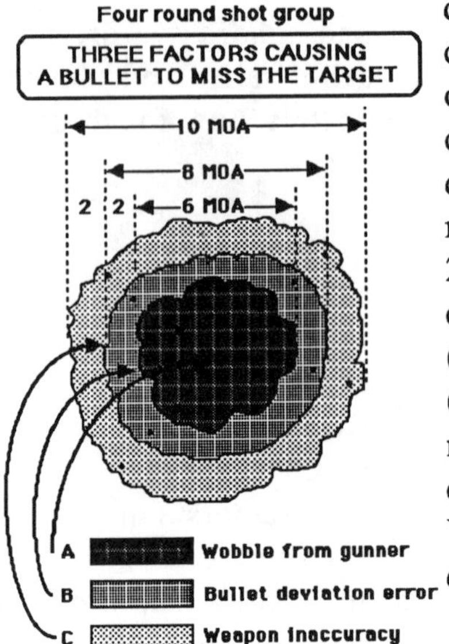

different source can either compound or cancel each other. Often it compounds; thus, the diagram. You raise the rifle and fire. At a scant 200 yards, your first round could be 12.6 inches (shooter) plus 2.1 inches (bullet) and 2.1 inches more (weapon) off. That could be a 16.8 inch miss. You might bring home dinner, but it's unlikely.

Drop into a sitting position, at which an average shooter wobble might be only three MOA. That will reduce your cone of fire to half size. Thus, you can figure out how far away you can shoot effectively from each of your own shooting positions.

> Distance in hundreds of yards times the MOA of any shooting position defines your cone of fire. Long distance multiplies wobble error. The farther you shoot, the wider your cone.

THE ALL-CRITICAL WOBBLE TEST

It's almost criminal not to test any military or law enforcement shooter before sending him or her into the field. Although they do fire on a range to get some idea of ability, military and law enforcement students

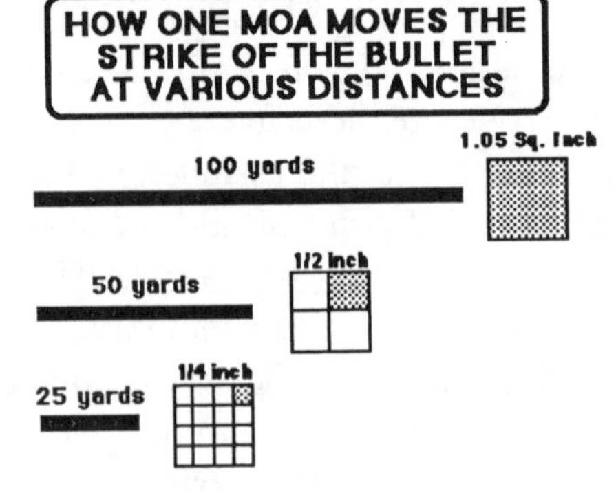

seldom learn to relate shooting accuracy to effective range. How important is it? Without this personal wobble information, you have no business shooting at long distance! Before going into combat, you **<u>must know</u>** the answer to this question. Learn about the size of your wobble and from that information determine your effective shooting distance. Ideally, aim at the grid and record your wobble from each shooting position at 25 yards. If you wobble across two grids, that's two MOA, a figure you arrive at by adding (1 + 1) the wobble on each side of the center bullseye.

> MOA wobble relates directly to shooting range. If you can't hold a tight group at 200 yards, you're better off not shooting at a target 800 yards away.

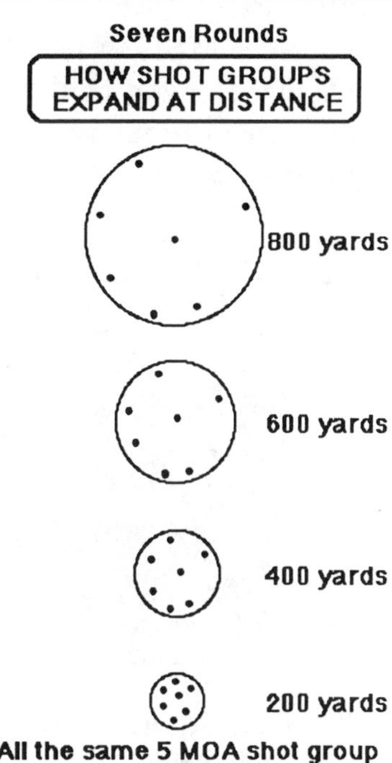

All the same 5 MOA shot group simply expanding at distance

Now, suppose you shoot from a position in which you, personally, produce four MOA. At 500 yards (5 football fields times 4 MOA = 20.25 inches) your cone of fire will be twenty and a quarter inches wide. Twenty inches may be OK for Bambi's daddy, but if you happen to be a Marine shooting at some enemy at that distance, you're probably going to miss and disclose your position—-especially at night. That's a terrible thing to do if you have NVG (Night Vision) capability and he doesn't, because your muzzle flash just told him where you are and made the conflict a 50/50 survival proposition. It will be worse if the enemy has many shooters with full auto capability; you'll get ventilated.

DISCOVERING NON-WOBBLE ERRORS

Once you find out how bad your shooting wobble is, test your ammo and weapon. Here we have some difficulty because with a bench rested weapon and ammo made by Mr Gun Genius, we can't find out for sure whether the gun or the ammo is causing the deviation. Tighten up your reloading procedures. Weigh each powder charge. Experiment with bullet seating depth. When your ammo is right and you still need a tighter group, work on the rifle.

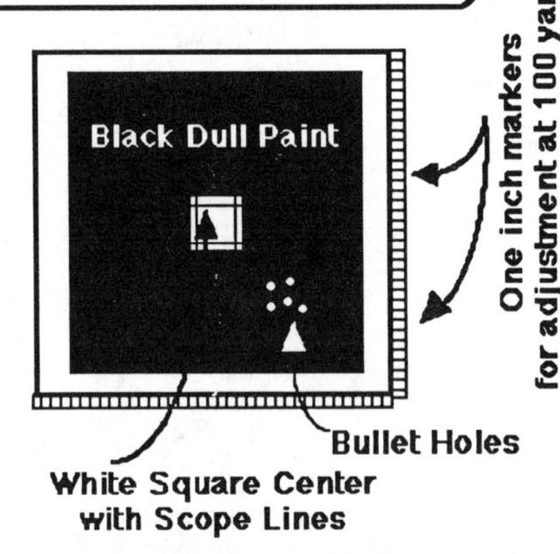

After you apply some of our magic improvement, fire-lapping the barrel, glass bedding the action, checking the crown, etc., guns will shoot tighter. If you use Hornady's match bullets seated to perfection and weigh every powder charge, the bullets should go through the same hole. Sometimes, though, they don't. The answer: Keep looking for the problem. Make up sets of five

TIGHTEN UP YOUR GROUP

Put your cross hairs on the lines inside the white square. Adjust four minutes right. Fire. Move the center of a good group into the sighting point after you fire by using the MOA markers on the side of your target.

cartridges each and code them to match a page of your rifle book (P. 123). Build our target with a white background and crossed lines on the square center bull. Don't try to hit the center; we want a group of shots in the black area so we can check for group tightness only. Try different powder loads. Change the bullet seating. Clean the weapon carefully. Sooner or later, you'll bring weapon and ammo into group-tightened heaven. You won't forget how to make that special ammo either, because we created a memory device to tie loading data to shooting results.

LEARNING SHOOTING POSITIONS FROM WOBBLE GRID

Finally, let's use "group tightener" on the shooter. You can sight on the grid and improve your position. When the weapon settles down as if bolted to concrete, you know which position to use and which improvements on that position work best. What might that mean? At one thousand yards (ten football fields end-to-end) you could place bullets consistently within a twenty-inch circle, which is the equivalent of two MOA. All you would have to do is compensate for gravity and wind drift, the two sole factors affecting bullet placement during exterior ballistics, (muzzle to target). That means all big game animals within a thousand yards of you will pay a personal visit to your freezer. It also means you stay out of harms way.

The grid is important because it teaches you which position works best for you. Try all three different sitting positions: Wide leg, crossed ankles and crossed legs. From which do you wobble most? While lying prone, check your wobble. Perhaps try cocking your right leg to relieve back muscle stress. Is your wobble less that way? Offhand shooters often can lessen wobble or tighten groups up by using a sling and making sure the weapon is supported equally by both arms. Try the sling and support the weapon more by pulling upward on the pistol grip with your shooting arm. Lock your rifle support arm down with a good loop sling, make a good shooting jacket, or use a rest. If you do all the above, then you cut down on shooting irregularity. With an accurate weapon, ammo home-weighed with match-grade, Moly-coated bullets, and some practice to reduce your wobble to one fifth of an MOA---you're good to go---a looong distance with a bullet.

> Look at how the size of your wobble changes as you change shooting positions. With a 2 MOA wobble from the prone it means you can reach out to 600 yards with only a 12 inch error factor. <u>The military needs to know this.</u>

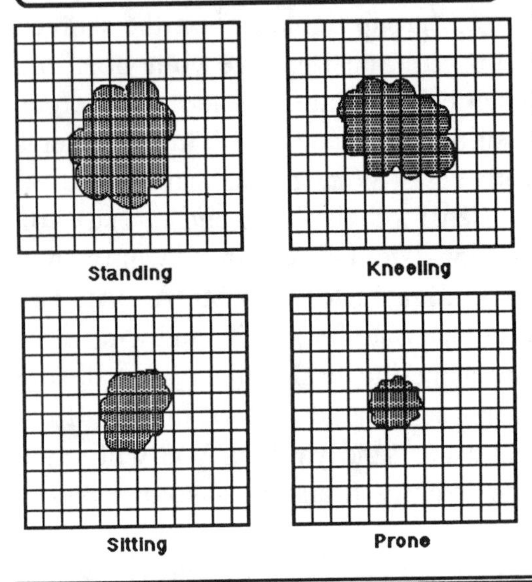

> Incidentally, look at the shape of the groups above. The standing guy is doing something to make his group vertical---could be breathing while shooting, or too much support on leading arm. The kneeler is vertical also, but is moving the weapon too much right and left. Most likely, elbow not directly under the rifle.

USING YOUR TESTED MOA TO DETERMINE HOW FAR YOU SHOULD SHOOT

From your MOA wobble

measurement, you know what position you **must use** to shoot at various distant targets. The average person, shooting from a standing position can probably hit a target from 100 yards because a seven-minute wobble would only cause a 7 inch deviation. Likewise, a sitting position (3 MOA) at 300 yards could contain a bullet in a nine-inch circle. For anything farther away than that, you would have to get in the position of prone-and-pray. When you can get your prone wobble down to less than two MOA, say 1.5, then you can plug the same nine-inch circle at 600 yards.

WOBBLE SHAPE

In addition to the *size* of your wobble, the grid reveals the shape of your wobble. Shape is important, because the shape relates to the kinds of error you make. For years, the military has used group shape (from shooting) and relegated it to certain causes. You can use the same mediocre method; just record your shot strings in the rifle book we designed for you.

As we already noted, a wobble from side to side is usually due to the shooter not placing his elbow directly beneath the weapon. Check your breathing as you watch your wobble. Some say you should take a breath, let it all the way out, then hold it. Others say half way out is good. Still others try not to breathe after taking in a full breath. I still hear arguments for one method or the other. Want to settle the arguments? Try both methods while checking your wobble on our graph.

Shooting only measures a group formed by shots which occurred at random **during wobble**. I've shot numerous times at a bullseye while wobbling all over, and the round broke while my sights were smack in the middle of the bull. Why? Trigger control. I applied pressure only while the sight picture was good. But the shot group would tell me I had no wobble. Therefore, I wouldn't be looking to correct my position, improve my balance, or change my sling use. On the other hand, our wobble grid measures shape from your widest deviations; therefore, it's a much more reliable measure. You discover errors you might never know about from examining the holes you put in targets.

An added advantage from our wobble grid is this: Say you adjust your position so you only wobble under one MOA. Then you go out and shoot a six-inch group at 100 yards. What? At least you know this error was not caused by shooter wobble. When your measured wobble doesn't coincide with your shot group, perhaps sight alignment was destroyed just as the round broke. Look for trigger jerk. Failing that, think about something really wrong with your weapon and ammo.

WOBBLE GRID AS TRIGGER CONTROL CHECK

With a laser sight on your rifle, you can dry fire and a coach at the target can see immediately if your trigger pull disturbs sight alignment. You can check this yourself if you watch the grid carefully through your scope (hi-power preferred). When you squeeze your trigger straight to the rear with steadily increasing pressure, your point of aim should remain as if cast in concrete.

SUMMARY OF WOBBLE GRID BENEFITS

Watch your wobble carefully. Your wobble defines the shooting distance from which you can take a shot that will hit. The shape of your wobble will also enable you to correct bad positioning habits. When you shoot groups which exceed your wobble, you can look for the cause in trigger error, irregular sighting habits, breathing during shooting, or rifling burn out. Finally, you can learn what positions work for you best. Best of all, you can learn all of the above without having to bust one primer.

This grid must be .2526 inch square each, so expand or contract on copier to make sure you have the correct size. Double it for 50 yards. At 100 yards, grids must be 1.05 inches each. A light or laser sight on your rifle will help a coach determine your MOA wobble. Note that you add values on either side, or top and bottom. If you wobble to the outer limits of 1, for example, your MOA wobble is **1+1 = 2.**

MOA GRID FOR 25 YARDS

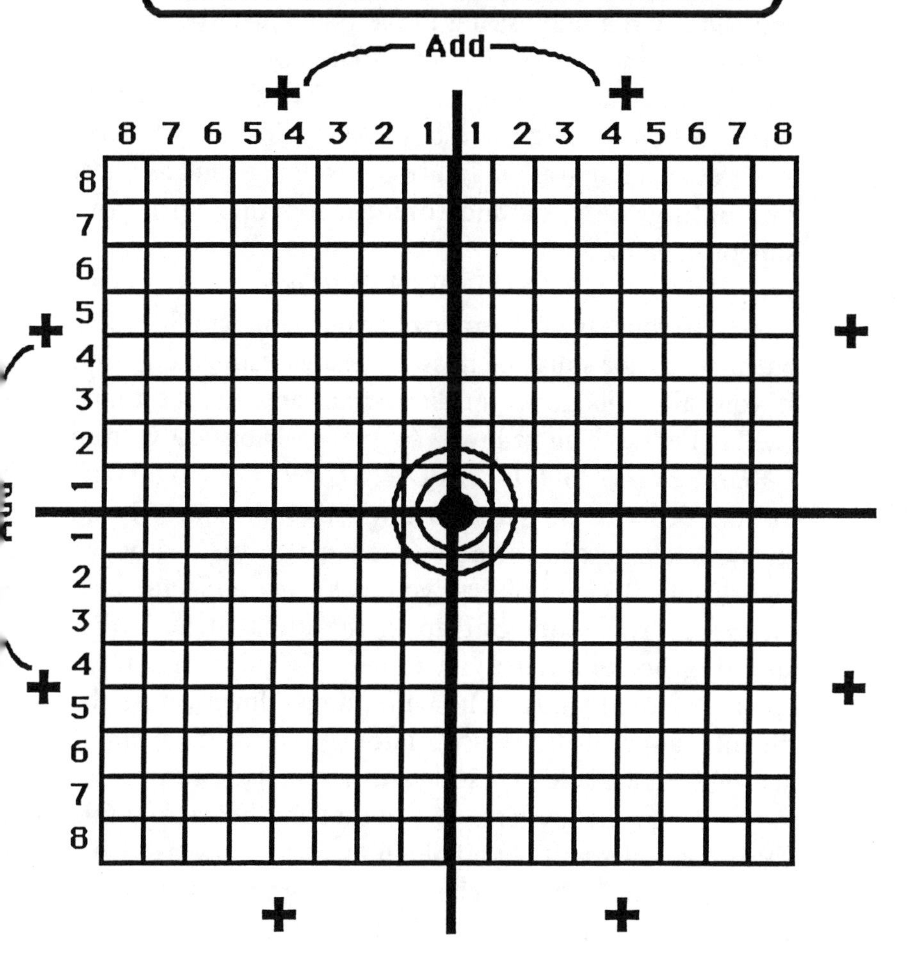

GETTING BETTER THAN GOOD WITH A REST

Things can be improved immensely if you use a rest for the weapon when you shoot. When your rifle's forearm rests on a solid object, you reduce your wobble. Most ranges today supply sandbags in which you can make a groove for your rifle and hold it extremely steady. Result: You can find your zero faster because a steady hold eliminates most causes of shooting error. Incidentally, you may want to have a sandbag with you in the field. Cut a pant leg off a pair of pants and sew one end shut. Put a draw string in the other, and you can use dirt from anywhere to fill it up. Better, sew both ends up after you fill it with kitty litter so it's light and easy to carry. When hunting use a 6' walking stick. If you have to shoot standing (offhand), use the walking stick as a rest by holding the stick and forearm together with your supporting arm.

UP AND DOWN ERRORS

If you haven't memorized your nomograph, watch out for a major cause of misses: Range estimation error. If you fail to judge target distance accurately, the round can land either high or low. As you probably know from ***EVERYBODY'S OUTDOOR SURVIVAL GUIDE,*** bullets drop at the same rate whether they are flying forward at thousands of miles per hour or dropped out of your hand. Because of that, you make any projectile land on target by shooting it up in the air a little. Most shooting books advise you to sight in your rifle for a couple hundred yards. The bullet drops down across the line of sight at that distance. In other words, the point of aim will have a hole in it 200 yards away. Hogwash!

I hate that system. If you do that, then at 500-600 yards the bullet will put a hole in the dirt. How bad is it? Well, with the .300 Winchester Magnum sizzling out of your muzzle at more than 2,800 fps. set to smack a target

200 yards away, the same bullet lands over a yard low at 500 yards. That's really poor. To fix the problem, you're supposed to aim above the target.

So you take your rifle hunting with a 200-yard zero. You see some delicious big game 500 yards away. Like the book said, you hold 46 inches above the animal and fire. Meanwhile, the deer steps forward while you're sighting above it, so your bullet is OK for elevation, but lands behind it---where it <u>was</u>.

How could that happen? If you have to hold over, you can't watch your target! Furthermore, we all know that range estimation in the woods is difficult. So how far is 46 inches when it's five hundred yards away? No-no---I have a better way to go, and since you bought this book, I'll let you in on it for free.

I want you to sight in for the maximum distance your rifle can deliver a bullet with sufficient energy. For a high power rifle, that's 500 or 600 yards. By doing that, you improve chances of hitting because:

A. You will hold under for shorter distances and thus

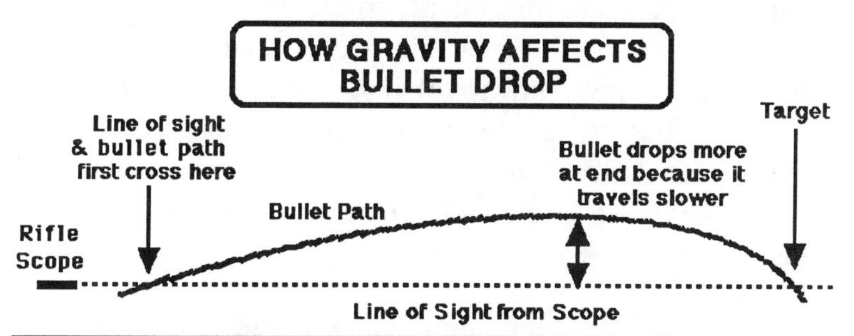

This is what bullet flight looks like in relation to your line of sight, which we show with the horizontal dotted line from scope to target. It crosses the line of sight on an upward path about 25 to 40 yards out. See Hornady---*RELOADING MANUALS---Volumes 1 & 2.* Since air resistance causes the bullet to slow down near the point of aim, gravity has more time to pull. Thus, more drop. Note double arrows. At that point, about two thirds of way out, bullet has highest distance over line of sight.

always have your target in sight instead of blocking it out.
B. You won't have to deal with as much adjustment. Even though a high power bullet will drop 46 inches under your line of sight at 500 yards, the mid-range trajectory of such a bullet is only 20 inches high.

Therefore, a mistake in range won't make much difference in target results. Suppose you're sighted in at 600 yards and you discover a target 300 yards away. Since you know from your nomograph that the mid-range trajectory of this particular bullet is 21 inches, you hold that much low. But the real range is 400 yards; you mis-guess-timated. At 400 yards, you should have held sixteen instead of twenty one. You'll be 5 inches low. Who cares? When you're that close to the point of aim with a rifle, only a bunny rabbit will be grateful for the error. All other big game will join your freezer club.

SHOOTING TIPS

Get some instruction. You can't learn basics properly from a book. Still, I won some gold as a shooter and Dave knows more than I, so we can help you. But I can't write everything in this book because Path Finder only wants new and novel information. A lot of the shooting basics out there have been tried and true for many years. Every instructor knows the basics. The best way to learn how to shoot is directly from a supervising instructor.

POSITION ADVANTAGE

Learn to sit down and aim quickly. I won a rifle match this way with a Garand once. I turned quickly to the side (right for me), crossed my legs and plopped. By doing that I narrowed my cone of fire because my sitting position causes much less wobble than my standing. At the same time I made the other "imaginary" shooter take on a target one-third normal height. That means he would have to be three times as good at estimating range and holding. Almost all shooters firing from a sitting position will win hands down against a standing shooter. Why? The sitting shooter has less wobble and offers a much smaller target to his enemy. The standing shooter has more wobble and offers his enemy a full-size target. With odds stacked in your favor this way, you'll win.

For all the above reasons, I don't like offhand shooting. The exception occurs with iron sights when you can lean your weapon's stock, not the barrel, against a tree or upon a rest. Another reason I don't like offhand shooting is the difference in eye relief scope shooters require. Eye relief measured while standing will cause some sitting shooters to catch a scope ring as it recoils into the eye brow.

That won't happen to you because you read this book. Set your eye relief on the scope so it's correct for your sitting position. If you have to shoot offhand, use a rest and crane your neck forward to get a proper sight picture.

WHAT ABOUT BRASS?

I should at least share the BRASS system with you. This is not mine. Somebody else thought this up a long time ago. I changed it a bit when I wrote ***EVERYBODY'S OUTDOOR SURVIVAL GUIDE*** so you could adapt it to modern firearms. As you read earlier, this is an acronym for Breathe, Relax, Aim, Squeeze and Surprise. You have to apply the word for a while until shooting properly becomes a good habit. If you fail to incorporate any one of these skills into every shot, you will burn a lot of gunpowder only to put holes in the air.

Some shooters don't know this: The BRASS sequence doesn't follow the order in which it's spelled. It really goes RABSS, but that doesn't spell anything we can remember easily. In practical order, however:

RELAX: Learn this; it's a vital skill. Tense muscles under any kind of strain increase your wobble. Make a conscious effort to relax your body. Begin with your legs, and think about relaxing every part right on up to your neck. As soon as you consider yourself to be relaxed,

AIM: Aiming through a scope requires that you to place your eye directly in the middle of the central-through-the-scope sighting path. If you see shadows on any side of the scope, your eye isn't centered. To aim properly with iron sights, you need to do two chores: Sight alignment and sight picture. They're two separate and distinct skills; learn them separately. First, align the sights. Second, place those aligned sights in proper relationship with the target to get a sight picture.

Sight alignment: Your rifle sights must be aligned perfectly. If they're not, then you multiply the error you created in 26 inches by hundreds. To achieve sight alignment with iron sights, you have to know the human eye can focus only on one spot at a time. Rifle sight alignment requires that the focus of the eye fixes on the following: The front sight, brought into alignment with the rear. Next, sharply on the rear sight without moving. Then your final focus goes back on your front sight to leave the rear sight blurry. Notice that "target" has nothing to do with sight alignment.

Sight picture: Now, once you have the sights perfectly aligned, you will place these in line with your out-of-focus target. To be a bad shot without possibility of correction, focus on the target. It's perhaps the most common error committed by poor marksmen. Once a shooter does that, both his front and rear sight will be blurry, which means he could be shooting with mis-aligned sights. That's death to accurate bullet placement. Suppose he allows his sights to veer out of line by as much as (1/32) one thirty-second of an inch, (1/32"). That error will be multiplied hundreds of times as the mis-directed bullet continues down range. When you truly hold your focus where it belongs just before you shoot—-on the front blade, then the target will appear blurry in the background. That's how you know you're shooting properly. The second the target appears well-defined in your eyesight, you can no longer keep your sights aligned (cause you're not focused on them).

BREATHE: Every time you shoot a rifle at a long distance, you need to hold it as still as possible. That's why you need to use a sling to make a solid position and place your elbow directly beneath the forearm. Even though you might not notice it with iron sights, breathing while shooting moves the weapon up and down. So stop. Take in a full breath, let it out almost all the way, and begin to squeeze with a slow steady pressure, <u>straight to the rear.</u>

SQUEEZE: The opposite of squeeze is jerk. You'll know right away what I mean if you've ever watched a novice take a picture with a camera. As soon as the photo is framed, the photographer stabs the shutter button, jerks the camera and causes the picture to blur. Do that with a rifle, even slightly, and the bullet you send down range will punch a hole through nothing but air.

To squeeze a trigger properly, you need a PhD. trigger finger which can increase tension smoothly and slowly on your trigger. Also, you can't push it off to either side; the trigger has to come straight to the rear. The best way to educate a trigger finger is to dry fire with a handgun. As you do this up against a white wall, you watch your sights. If you succeed in pulling straight to the rear, the sights will remain in place. Of course, our wobble grid does the same. Use a laser dot, and check to see if it moves when the trigger snaps.

HOW TO TIE LOADING & SHOOTING DATA TOGETHER

You may copy the work sheets in this book and reduce them to stick on ammunition boxes. Also, copy them slightly enlarged to 5.5 x 8.5 on both sides of the paper, and make 50 or so, then have them spiral bound.

Code ammo and page together. Color highlight through the Title Block. Use a penned-in letter designation A thru Z for each color. You may also have A sub 1, 2, etc.

On line a, record the kind of rifle, caliber, shooting range to target and mid-range. Line b is for muzzle velocity (Hornady Manual), etc. Line c deals with weather conditions. On line d, record the light conditions and the nomographic data from Hornady. On separate graph paper, build a nomograph for this round, and don't forget to write down how much energy you deliver to target.

Shoot four groups. Recording your elevation and windage adjustments for each group lets you check MOA with bullet strike from the center of each group.

At k, note the wind and your estimate of strength. In the wind value box, write down how many MOA clicks you assigned for this wind on the left side. Directly under that, write down your corrected sight adjustment after your first few rounds tell you how much you misjudged the wind. The second set of assigned and corrected figures can be used if you get a wind change---in direction or strength.

Above the target, m, date and time refer to when you loaded this lot, which helps you trace severe error to cause and perhaps would be the basis of a decision to trash other rounds.

The comment box has space for loading and performance notes. Loading might be bullet seating data, sizing data (full-length resize or neck size only). Performance notes would be deviation from zero elevation (traceable to humidity or altitude) and mistakes on certain rounds you may have made while shooting---such as trigger jerk. Number the rounds on target by sequence. If a bee stings you on the nose just as you shoot #4 in a certain lot, you probably can disregard that performance.

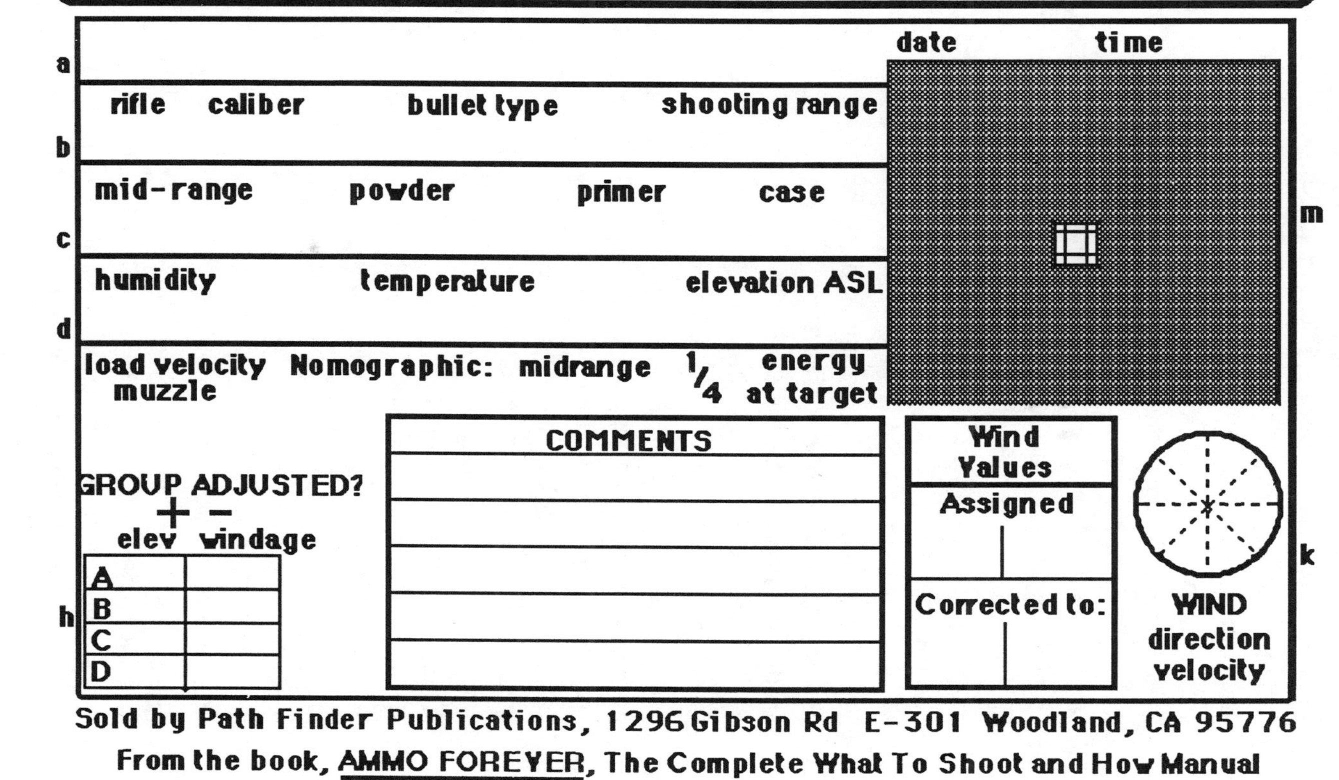

RIFLE CARTRIDGE: LOADING & SHOOTING DATA

a

date time

b

rifle caliber bullet type shooting range

c

mid-range powder primer case

d

humidity temperature elevation ASL

load velocity muzzle Nomographic: midrange 1/4 energy at target

m

COMMENTS

GROUP ADJUSTED?

\+ −

	elev	windage
A		
B		
C		
D		

h

Wind Values

Assigned

Corrected to:

k

WIND direction velocity

Sold by Path Finder Publications, 1296 Gibson Rd E-301 Woodland, CA 95776

From the book, AMMO FOREVER, The Complete What To Shoot and How Manual

SURPRISE: is the only military word Gomer Pyle ever got right. As a shooter, if you don't surprise yourself when the round goes off, you'll be surprised at the lack of holes in your target. Why? When you know the round is about to break, your body reacts. Even subconsciously with the smallest of reactions, the weapon moves. The result will show on or perhaps off, target. You may actually jerk the trigger. I've seen some offhand shooters hunch the weapon just as it fires so the bullet dumps low in the dirt. Therefore, this last element of a good shot is important. Be patient. Wait as you increase pressure on the trigger.

TEAM BASIC SHOT GROUP ANALYSIS

Long-distance shooting is done best with a two-man team. One member uses a spotting scope and reports to the other where the bullets land. Consider the bullseye as a clock. Twelve is on top, six on the bottom. Use the value rings on the target to designate the distance away from the bull. So a "four out at one o'clock," tells you the round went into the four ring, high above the bullseye, just out to the right a bit. Later when you're shooting at live targets, your spotter will use the same system. Many years ago, I used to pop rabbits in the evening on a golf course with a .22 long rifle out of a Marlin lever action. The wind coming in off the sea would cause the bullet to drift, but my spotter would tell me, "Three feet off at four o'clock." A quick adjustment would provide dinner.

TEAM SHOOTING

I'm bothered big time by the U.S. military strategists who fail to provide shooting overlap to the troops. Just as you overlap optics, do the same with bullet launchers. One should be a scoped, heavy bullet, long distance (.30-06 or magnum) for accurate placement. The other should be a .223 (military) with high magazine capacity. Both cover fields of fire at different ranges.

Close range shooters can use open sights---but need rapid fire capability.

On hunting trips, two of us normally split up and hunt assigned sectors. We also use a drive-ambush system. One sits and the other crashes through the woods and tries to shove everything toward the stump sitter.

Sometimes we hunt apart but on the same azimuth, changing at pre-arranged or hand-signalled times. In such a case, one of the two man team is designated driver, and he decides which way to go. In other instances, either one of the two can follow tracks and give the direction changes to the other.

For military purposes, a sniper team stays somewhat together because they need to communicate. One uses binoculars and spotting scope. He finds targets, defines them and reads the wind. The shooter's job is bullet placement after taking windage data from the spotter. If the team gets into trouble, both could shoot. I don't like that system.

In my opinion, two snipers working as a pair would do much better to stay apart. All you have to do is solve the commo problem. How? With one earphone and a swing mouthpiece to a walkie-talkie type radio. That way, you discover more because of the two viewing angles you have of any hiding place. Because you outflank any enemy who hides behind a log, he's is exposed to either yours or your spotter's vantage point. You get highly effective suppression fire since both snipers can shoot. After engaging, you can rendezvous at any predetermined point, or you can cross paths on time for a meet with SOCKNAV vectors, (See ***GREEN BERET'S COMPASS COURSE***).

The same tactics apply to defense. Your security is enhanced when you stay a distance apart. In the U.S. the vast majority of criminals carry handguns. Two riflemen apart can shoot near the other's position and overcome almost any attack. Cover your approach with bird seed so early morning intruders make a flutter. Once awakened, one rifle shooter always will be out of handgun range and thus, can cover the other with ease.

SHOOTING PRACTICE SOLO

Long-distance shooting is preferred because you can check your zero, learn about wind and gain confidence at long range. On close targets (not paper) you often don't see where the bullet landed because you're still in recoil. In desert, a puff of dust usually tells you where

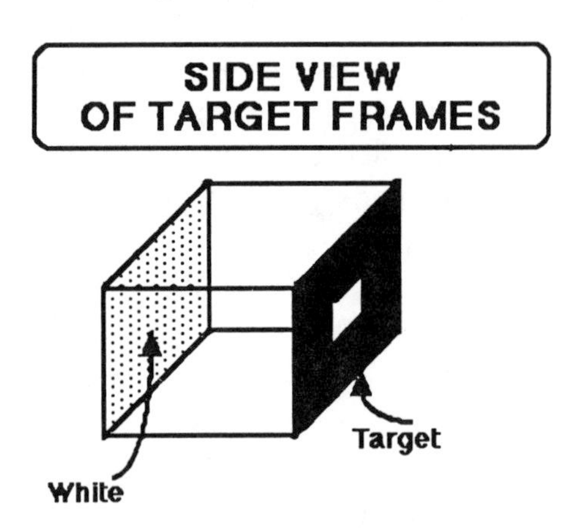

the round landed. Fire and return to your sight picture as soon as possible. On paper you need optical help. A white background behind a black target makes seeing your bullet holes much easier.

Build a cubical target frame. On one side, use black paper (paint) and the other side white. Punch a series of bullet holes in the paper and then move off to one side; with a white background behind, the holes will show up clearly.

SHOT GROUPS—-TARGET PRACTICE

Examine your shot group and try to discover a pattern indicating errors. Shots up and down show a rise and fall in the muzzle. Breathing in and out while shooting is probably the cause. Shots in a horizontal line left and right mean the muzzle was wobbling in that direction. Why? Probably, you didn't have your elbow under the weapon and muscle tension—-contractions and relaxations—-were causing that movement. Shots down and to the left grouped at 7-8 o'clock (right handed shooter) suggest trigger jerks. Shots consistently out to the left at nine o'clock probably mean the trigger finger is pushing against the stock of the rifle.

SIGHTS ADJUSTMENT—-IRON

Knowing your sights are zeroed correctly is all-important. You need confidence in the weapon to shoot well. Often, iron sights get knocked out of left/right adjustment, so make sure they're right. You'll be using a soft brass drift punch and a hammer to tap them into alignment. To do that, drop a little dry dirt or desert sand on your rifle, then blow it off to leave a fine coat of dust. Make two aligned marks in that dust and you will know how far you move the sights when you tap them with a (soft brass) drift punch. No dust? Mark the sight and barrel with a knife. The idea is to make sure your rifle shoots exactly where you aim, neither left nor right. With sights set dead ahead, aim into a cross wind to compensate for wind drift. Raise those iron sights to produce a high shot at 150 yards, which goes in right over the blade at 300 yards. You're ready to hunt.

TACTICAL LIGHTING

If you hunt, you'll be operating during the day. In the defense mode, learn to shoot at night. In either case, tactical lighting is important. Basically, shoot from a darker area into one with more light. Facial hair is a plus in the woods. Color yours green if you look like Kenny Rogers. Since we now know deer can see ultra violet light, use special soaps to cut down ultra-violet light when you wash your hunting clothes. During the day, that means stay in the shadows, and shoot into the light. In late afternoon hunt the North slopes because the sun will light up the hillsides exposed to the South.

At night, wear no bright clothing. Shoot from shadows into street light, moonlight or any other source available. Tactical lighting doesn't have to be absolute; you don't have to be in pitch black with your enemy in the spot light. Just make sure you're relatively in the dark. You can change the tactical lighting in your favor if you use any kind of trap with explosive. Say, for example, that your enemy has to come through a certain area of the woods at night. A trip wire which sets off a loud flash not only alerts you to his presence, but will destroy his night vision for half an hour—-plenty of time during which you can see and he can't.

CAMOUFLAGE

Don't be seen; don't make a sound. You can learn to do all this, but it takes time and effort. For hunters, know that you are dealing with an animal with a superior ability to hear. Test your boots. Use a tape recorder with a sensitive pick up microphone. Drop the mike down near your boots and walk with them. Listen to the tape. Are you a one-man band in these boots? If you already own them, oil them with castor oil to make the leather soft. If in a store, buy a pair that won't alarm the whole forest.

Scent. For hunters, clean your body with no-scent soap. You don't want animals to smell you. For warfare people the problem isn't scent; it's movement. You will move too much if you let creepy crawlers, mosquitoes and gnats use your body for a feeding ground. One solution: Add (cider) vinegar to your water intake. You'll stink so bad no bug could love you. Others: Blouse your boots. Use insect repellent on all exposed areas.

Slow movement. Finally, learn to move slowly while stalking. Be deliberate. Never stop unless someone is staring at you. Never move fast. If you turn your head, turn it in slow motion. Present your human form to the woods in which you operate as something nobody would recognize as human. Thus, don't stick up head with hat on it straight up. Tilt your head over toward your shoulder and raise up so that one ear, then one eye slooowly comes up above your concealment. Television has unwittingly set the standard by which most observation takes place: Sixty frames per second. If you learn to move more slowly, even while someone is watching you, it's extremely difficult for them to comprehend the reality of your movement.

SECURITY

Don't forget security when you practice shooting. On ranges you don't have to worry much, but if you drive into a remote area to shoot, be careful. As it happens now in the United States, a lot of bad guys are out there in the woods practicing also. The first bang out of your rifle could draw a gang from miles away. Most have optics with them, and they'll scope you out if they get a chance. A solo rifle shooter could easily become a victim.

Position yourself so there is only one avenue of approach toward you and you can observe it. Get up high. Don't position yourself and your targets down in a rock pit where someone could walk up to the edge of a hill and look down on you. Never leave your weapon on a firing line; it stays with you always, loaded. Some weapons thieves have been known to wait for you to quit firing, then attack. If you plan to shoot 100 rounds, bring 200. Don't go home without ammo. There is no point in owning a weapon for defense if you don't have ammunition for it.

SUMMARY

Learning to "use" a rifle requires patience and never-ending-improvement. Once you know your rifle shoots perfectly and the ammunition you feed it is precisely consistent, the only factor you have to work on is your own shooting ability. To reach a high degree of skill, you need to learn to shoot tight groups from reliable body positions.

With your wobble measured and reduced through practice, other shooting skills will be much easier to isolate and make perfect. Then you'll be able to place a bullet with precision on a target so far away it will never see or hear you until it becomes holey.

Always shoot from a distance close enough to insure hits within your measured wobble-ability. With practice, you can be the master of any area in which you choose to operate.

THREE KINDS OF BALLISTICS
INTERIOR + EXTERIOR + TERMINAL
All play an important part in efficient bullet operation.

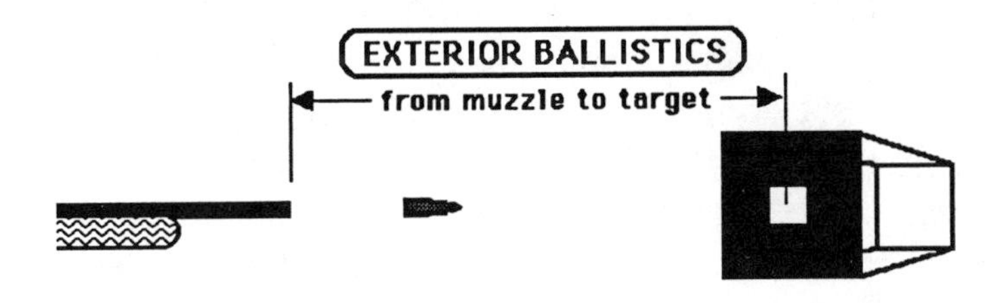

III Chapter 4

EXTERIOR & TERMINAL RIFLE BALLISTICS

Exterior ballistics (between muzzle and target) pertain mostly to rifle bullets. We shoot handguns at short range so we don't pay much attention to exteriors. Perhaps small caliber, light weight bullets will drift left or right in a good wind out of a handgun, but we're not too concerned. Shotshell exterior ballistics come into play because smaller shot slows down easily and therefore has little effect on most targets at long range. When you increase the shot size, you reach farther with more effect.

So exterior ballistics are about rifle bullet flight **after** it leaves your muzzle and **before** it hits the target. Knowing your bullet's exterior ballistics---the effect the wind will have and the nomographed performance---is all important for scoring from a distance. Once you start reloading for any weapon you've improved, you can use exterior ballistics as an aid to make your rifle perform long distance magic.

COMMERCIAL EXTERIOR BALLISTICS

Following is a list of the best long range cartridges on the commercial market. We list these in the order of their flat shooting ability out to 500 yards while still having 1000 ft. lbs. of energy. These require much powder, create high pressures, but deliver only 100 to 200 fps. more velocity. They destroy brass and burn out barrel throats faster than standard long range cartridges. We don't recommend them because a survivalist needs longevity more than power.

Cartridge	Bullet weight	Muzzle velocity
1. 6.5x68 RWS	93 Gr.	3770 fps.
2. 264 Win.	100 Gr.	3600 "
3. 257 Wby. Mag.	100 Gr.	3600 "
4. 7mm Wby. Mag.	140 Gr.	3225 "
5. 300 Wby. Mag.	178 Gr.	3120 "

If you shoot these calibers, down-load them to save brass and rifle wear, especially in your barrel, where hot loads burn out rifling and barrel throat. For hunting purposes, you can load to the limit, but you don't have to shoot many bullets when hunting. Down loading helps your rifle to last years longer.

REDUCED LOAD CAUTION

Some loads with less than 20% of suggested powder have auto-detonated. Don't an unexpected explosion by reducing powder less than 20%. The majority of experts believe this phenomenon only occurs with magnum or over-bore capacity cases in which half charges of slow burning powder and weak primers are used. As a side note, think about any round you put together blasting off all by itself (auto-detonating). Be careful where you point that thing! Make others be careful, too. The possibility of auto detonation is the major reason you tow your rifle in a drag bag, pointing behind you In the military mode, I would instruct you: Do not alter God's plan. You were issued with only one hole in your posterior.

From the muzzle to the target, **only two factors** have an influence on bullet flight-gravity and wind from the side. Wind resistance plays a part, but that's only because some projectiles are shaped so they de-accelerate rapidly—-thus giving gravity more of a chance to pull. Translation to the shooter: The bullet slows down a lot and has time to drop more on the way to the target. Bullets with different coefficients can cause the strike to be severe. Five hundred yards away from the muzzle of a .308 rifle, two different bullets of the same weight will strike over 14 inches apart. Why? Bullet shape. On those same two bullets, full vector wind drift (from the side) at 10 miles per hour makes a difference of over 16 inches.

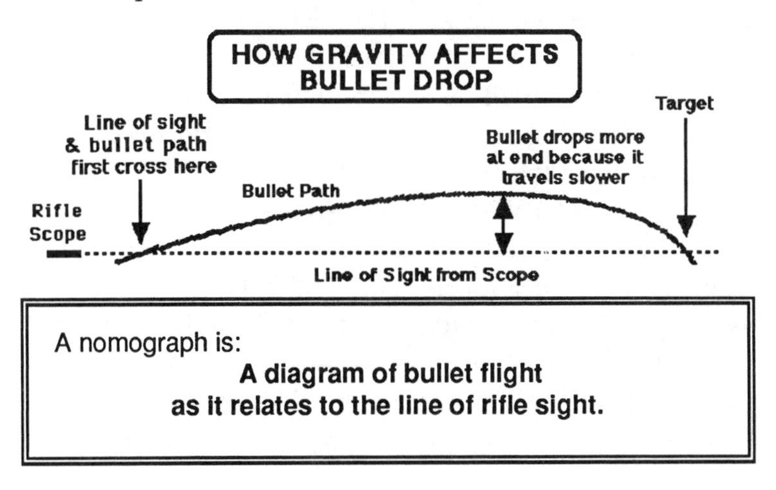

A nomograph is:

A diagram of bullet flight as it relates to the line of rifle sight.

YOUR PERSONAL NOMOGRAPH

To be a safe and competent rifleman, you need to know both yours and your rifle's potential. With that combined knowledge you'll have a practical idea of how far you can shoot, hit and kill a game animal. Discover your personal potential on our MOA grid target. Assuming you can hold your sights tight enough to shoot a small shot group at a distance, your only problem will be vertical elevation. How will your bullet fly? Every rifle bullet travels upward, over the line of sight, and then curves back down to land on target. We describe that curve with a nomograph.

A line of sight goes from the center of your eye through your scope and extends straight beyond for hundreds of yards. That straight line (of sight) is the reference point for bullet flight. It tells how a bullet flies up and drops down on target. To discover that, use graph paper and pen to construct your personal nomograph. Memorize it. Then, either read ***EVERYBODY'S KNIFE BIBLE*** or carry a range finder to estimate range accurately. Once you have zeroed your rifle to strike on the point of aim from a certain distance, you'll know at any shorter distance exactly how far above the point of aim your bullet will strike. Therefore, if your zero lands on the point of aim at 500 yards, the highest point above your line of sight will be at approximately 300 yards. If you get a short shot, hold your point of aim lower than where you want the bullet to strike.

HOW MUCH MID-RANGE?

Although you can use reloading data from other manuals, I only use one—that's from Hornady. (I also use their bullets). That manual shows the position of your bullet above the line of sight in 100 yard increments, and you can accurately compute the differences between. Make your line of sight a level line across the paper. At just beyond mid-range, plot one point above the line of sight. Use the figures for your bullet in the manual. Do the same for other points—-100, 200, 300, and 400. Now, connect the lines with the french curve, noting that the farther away from your muzzle, the faster a bullet drops.

Note: Some Hornady bullets fly better than others. Those are the ones which have a better aerodynamic (cigar-like) shape. Called "boattails" in the trade, they slide through the air more easily than other bullets. To make them fly best, Moly-coat them with a kit from NECO. Look at the bullet coefficients in the manual. The higher the number, the better. My favorite is the Hornady 7 mm. Remington Magnum; it's over 500. Compared to a .308 Spitzer down around 300, you get almost twice the long distance performance.

Remember: Bullets slow down faster in heavy air. High humidity, cold temperatures, low altitudes and low barometric pressure slow the bullet down. Therefore, gravity gets more of a chance. To counteract that, hold higher---about one MOA. The reverse is true for high altitude, hot temperatures, and dry Arizona air. Hold lower; your bullet is going to sizzle all the way to the target; therefore gravity won't have as much time to pull it down. If you like number crunching, the Hornady Manual provides the formulas. You simply plug the figures in and do the math to get an exact change for your bullet coefficient and trajectory. Otherwise, one MOA will be close enough.

WIND DRIFT

Winds from the side blow a long range bullet off target. How much? Longer, lighter bullets move a lot. Shorter, heavier bullets aren't bothered as much.

When I competed for 6th Army, an older Sergeant taught me what to do in a long range match during a heavy wind. As you may know, you get twenty minutes to shoot twenty rounds. So the clock starts ticking. Nobody wants to shoot because all the other shooters will be watching the marker to find out how badly the wind is moving bullets around. Also, heavy winds often blow a round completely off target. That's bad news. Not only do you score Maggie's drawers, (a miss designated by a waving red flag) but you don't have a bullet hole marked anywhere to use for sight adjustment. So I shot a deliberate miss; I dumped a round between two targets. As luck would have it, I put a hole in my target instead of my neighbor's. Once I had a hole in paper, I could adjust fire. So I moved my sights to compensate for the wind.

WIND. HOW MUCH VALUE?

How many MOA you move the sight depends on where the wind is coming from. Give winds coming from the side full value. Winds coming at you from a quarter direction get half value, and you compute for directions in between. Watch your first shot carefully; you'll be using the strike of that bullet for future adjustment.

Turn your spotting scope into the wind until you get no sideward wind wave across your scope; it will appear as a boil; little curving lines going straight up. When you see a boil, the scope points toward the wind's direction. Another way—-construct a telltale from a piece of yarn. Hang it on a branch of a nearby tree or a stake near you so you can see exactly from where the wind is coming. You'll know also when it changes direction. In the field, I like both methods. If you have the time to play with the scope, you can find the boil and adjust. On cloudy days, however, you may not be able to see the heat waves, so the telltale is invaluable. Also, you get immediate wind change information from this little piece of yarn. Tip: If you're tactical, hang it out of sight of enemy sniper optics.

Note how the direction of wind across the bullet's line of travel will affect flight. Quarter value and three quarter value winds come from between half and full. In our rifle book, the wind value circle corresponds. Keep good records. Confucious says, "Shooter with faintest pencil mark in Rifle Book better than hand loader with strongest memory."

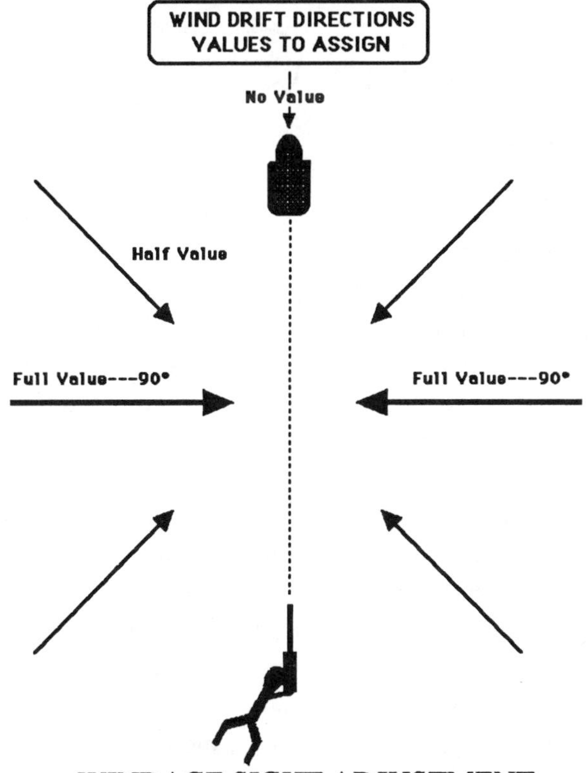

WINDAGE SIGHT ADJUSTMENT

With iron sights, you move the rear sight <u>into</u> the wind. Each MOA click will move the strike of the bullet in the target 6.3 inches at 600 yards. Scope adjustments vary. Read the scope maker's booklet so you know exactly how many clicks it takes to adjust one MOA, Minute Of Angle.

TERMINAL BALLISTICS

People like Hornady have spent complete lives studying this subject---how a bullet acts and performs after it hits the target. Though you shoot at something you can hit far away, sometimes your target doesn't become freezer meat. Why? Your projectile doesn't have enough energy to work its magic at that distance. Some slow down because they don't weigh enough. Others don't expand to cause a killing wound.

Since most shots are taken late in the day, terminal ballistic failure means you wind up tracking until dark, after which you pretty much have to sleep out in the woods and begin looking again at first light. I've talked to some hunters who lost the track. With the animal lying quietly in the brush after sundown, a hunter could be a few feet away and not discover it.

Without a perfect bullet, you get clean kills in one of two ways. A. Shoot a cartridge so heavy you have to load it with a crane. Push it with more gun powder than they use on the 4th of July. Don't flinch from the shot and land that bullet in a vital area. B. Shoot a lightweight, speedy bullet into the central nervous system. I've interviewed game wardens who swear by .243's because they could place the round up behind the animal's ear. I've done that, also. Once with a 6 mm. round and open sights at 167 steps away, my deer went down hard immediately.

Notice how neither of the above might not happen. A light weight bullet poorly placed will only wound. A heavy bullet aimed well might easily miss because heavy bullet trajectory curves are severe. To shoot accurately, you'll have to know your nomographic curve precisely and estimate the shooting distance exactly.

For example, lets consider the .308 Winchester 180 grain bullet. If you sight your rifle to shoot 2.5 inches high at 100 yards, what will be the bullet's path of trajectory? At 200 yards you will be 0.2 inches low with 1896 ft.lbs. energy. At 400 yards you will be 28.5 inches low (under your point of aim) with only 1269 ft.lbs. energy remaining.

For good terminal ballistic effect, you need two things: Penetration and expansion in the target. Although many bullet makers produce a variety of results, this is how Hornady makes great terminal ballistics happen. They control the expansion and performance of the bullet by varying the thickness of the copper-zinc jacket composition. Thin jackets almost explode on contact; thick ones penetrate. Perhaps you get the best of both worlds with tapered jackets, which expand rapidly during early penetration, and expand slowly as the bullet penetrates deeper into the target. With an internal interlock ring, they marry the lead core and the jacket together. Result: Deeper penetration with good expansion.

In Volume I of *HORNADY'S HANDBOOK OF CARTRIDGE RELOADING,* they list the 7 mm. BTHP (Boat Tail Hollow Point) in 162 grains with a ballistic coefficient of .534 as being for match use only. I don't agree, but I respect them highly. The important thing to note in the tables, however, is the velocity level at which the bullet is efficient: 2,300 - 3,200 fps.. Now look at Volume 2. You can hit with 2,225 fps. at 500 yards and smack with 1,781 lbs. of energy at that distance. Trajectory is only 14.8" above the line of sight at 300 yards. That means you could get excited during the hunt and forget to hold under, but a strike of the bullet 14.8" high will still be a solid hit.

Build a round to deliver enough smacking energy at your preferred shooting (long) distance. The farther you shoot, the more power you need. Once you have the power, build a nomograph so you can hit with exactness at any range shorter than your max. Thus, no matter where you encounter any target, you'll hit with efficiency.

III-Chapter 5

DEALING WITH INCOMING-RIFLE

The whole world is at unrest. Path Finder sells books in areas where bullets fly daily, such as Israel, Africa, and Los Angeles. Of course everyone knows that America is the most violent nation in the world. Translation: Someone may shoot at you. How not to get shot at and hit is an important topic for any book of this nature. We think others leave it out because it's not fun to read.

AGAINST A SHOTGUN

While carrying your rifle, if you get shot at with a shotgun, two possibilities exist. First, you allowed someone to get too close to you. Assuming the first incoming round misses, duck. Get down, out of the way by hitting the dirt or dive for cover. If the shotgunner looking for you isn't close, then he has to be one of the dumbest people in the world. Anybody who engages a rifleman at a long distance (150+ yards) with only a shotgun can't be too bright.

The best he can do in range with any shotgun is 125 yards. If you can open the range to a couple hundred or more, it will be absolutely no contest. On the other hand, don't engage a shotgunner at his range unless you can punch through his barricade and surprise him. Above all, don't expose yourself. Shotgun patterns spread out and cover a wide area; if you happen to be within that cone of shot, you'll be hit. If that happens, all you can hope for is small pellets (which lose their kinetic energy quickly).

AGAINST A HANDGUN

Most incoming problems will be from pistols. That's because most of the shooters in the country are criminals and they carry concealed weapons. Also, criminals aren't too bright as you may have noticed, so they learn a lot of their weapons tactics from the movies. In old Hollywood, six shooters could blast away with over twenty rounds without a reload. In new Hollywood, Charles Bronson takes on a whole gang of thugs and wins with a .32 revolver. As recently as a couple of years back I saw Steven Seagal run down a corridor of a battleship spraying 9 mm. shells out of two crossed automatics.

If a handgun shooter engages a rifleman at a distance, his elevator doesn't go all the way up to the top. Once again, the range must be short for the handgunner to do any damage to you. At distances of over 100 yards, the average handgunner sprays bullets all over the landscape while the rifleman can often hit a nickel. Perhaps the exception might be a shooter using a good rest with a long pistol barrel and a scope and a hot load out of a .357. That might get to you. Normally, however, to hurt you the average handgunner needs to close range. Getting closer to a rifleman is not a good idea.

If an incoming handgun round misses you, it probably came from close range. Stay down; stay safe.

Think about the noise the handgun made---big boom or little pop. You may be safe behind a barricade. If in law enforcement, call for help; the cavalry should show up pretty quick. Perhaps a SWAT team member is close by. He can come in on the perp's position from the side and one of you should be able to send a *sayonara.*

Often a few seconds is all it takes for the handgunner's second or third round to find you. Remember, he is probably shooting some kind of 9 mm. autoloader. That means it's an incoming round weighing around 110 grains out of his muzzle at around 1400 fps. Those rounds can hurt. The reason for the first miss is probably caused by his shooting across too much distance. He never heard of our wobble graph, so he'll be taking shots only John Wayne could have made on a clear day. However, the key word is "autoloader." Even if his first round doesn't score, it will be followed by several more. You have to take cover. Don't expose yourself trying to shoot back until you know you're out of danger. Once safe, you can't get hurt; relax. Then---start solving the problem. A typical law enforcement scenario is: Cop near car gets incoming. With a rifle, even as small as a .22, drop down on the ground and shoot under the vehicles. Even with bullet drop, you'll get a nicely placed ricochet.

AGAINST ANOTHER RIFLE

Now you have a problem. This is the occasion which could cause you real trouble. Don't shoot back. Most snipers have concealed themselves well behind good cover, and a return round won't phase them. But it will leave you exposed, even though you have dropped to the ground and reduced your target size substantially. Could he hit you? I think so. He is most likely using a decent rifle with a rest, which means he can shoot under two MOA. At six hundred yards, that means a 12 inch group.

You could be in serious trouble. Take cover and change position behind that cover immediately. Most riflemen will try to penetrate the cover. Once you're out of harm's way, you can solve the problem.

If you can, determine what kind of rifle is shooting at you. Almost all enemy soldiers shoot the same kind of weapon, and the sound that weapon makes is distinct in two ways—-the pitch of the boom and the cyclic rate, if automatic.

First, estimate his distance away. Almost all rifle bullets will be supersonic when they whistle past you. Want to hear one? Go out into the desert or woods and duck down behind some good cover; then have a friend shoot over you. The bullet will "crack" as it passes. Right after the crack, you will hear the boom from the weapon. The difference in time between the crack and the boom tells you how far away the shooter is. You tell that time in two ways. A stopwatch is most accurate. At the crack, punch it. When you hear the boom punch it again and read the elapsed time. Less accurate but easier to do is the counting system. Look at your LED display watch as it flies through time, preferably in stop watch mode, and teach yourself to count quickly, "one, two, three, four, five," in exactly one (1) second.

If you hear a crack pass you, thank God later, count now. Count to five with that quick cadence until you hear a boom, then stop. Ask, "What number were you on?" Let's suppose you had just counted, "three." Three fifths of a second is 6 tenths, or 60%. How do they measure the speed of a bullet? In fps—-feet per second. How fast did the bullet aimed at you come out of his muzzle? Depends on the gun, but you can tell that from a variety of intelligence input. Foreign army? Probably AK-47, which flies along at just under 2,000 fps. So—-if an AK round is looking for you and it took 60% of one second to come close, you multiply 60% times 2,000 feet to get 1,200 feet, or 400 yards.

Let me digress. Be careful when you move. Though an AK shooter is too far away to shoot accurately, you may have guessed wrong. He could be shooting a 30-06 with mix 'n match capability. A hotter round may follow. Also, his weapon may have been worked on by a gunsmith who knew his business. Don't expose anything you don't want holey.

Now, back to the main problem at hand. Where is this illegitimate child? We know at least 400 yards away. What direction? If the bullet hit anything near you, it left a hole. When you place a locator index—-a pencil, cleaning rod or decently straight twig—-in that hole, it will give you the direction the bullet was traveling. The opposite way will give you the shooter location. Once you know the line and the distance, the shooter will have a problem.

> Of course, you'll ask, "But how do I place something in that bullet hole?" Good question. The answer is, "Very carefully." If you happen to be on a family outing when you get shot at, tell your mother-in-law to do it.

Failing that, you will have to obfuscate the area. Smoke is a good idea if you have a smoke grenade. Dust is another; simply kick up a lot of it by throwing handsful of dry dirt in the air. With superior fire power you might pepper the general area with concentrated fire while someone places your locator index. That's risky, however, so be quick. Once you have the shooter located, plan. What do you have available? Can you out-range him? Can you get help and out-flank him? If you can see around your barricade, you can probably get off a good shot of your own. After all, you know the distance and from looking at the bullet hole you might get a decent idea of the direction.

Incoming is a problem only if you are within range of incoming weapons. Once you get some distance between you and the other shooter, though, you and your rifle should be able to solve the problem easily.

It burns me that only good guys are prosecuted for carrying guns. The criminals in our society are never bothered with the problem. Why? Bad guys always get caught after a higher profile crime, like a double murder. The prosecutor then tries to prepare an airtight, irreversible case. You never hear of a prosecutor charging the criminal with the murders, thefts or whatever and then saying, "Oh incidentally, he used an illegal gun, your honor." On the other hand, a police officer will ask someone quite innocently if he or she has a gun. If a good citizen answers truthfully and says, "Yes," they confiscate the weapon and press charges on the only crime they have to deal with.

III-Chapter 6

CONSTRUCTING RIFLE AMMUNITION

Please note a few definitions:

Stroke *verb.* It means to grasp the press handle, push down or pull up and make the ram move.

Press handle The long pipe-like device sticking out from the press, usually with a handlebar grip on it.

Ram *noun or verb.* The noun definition refers to the part of the press which moves toward the die when you activate the press handle. You can "ram" *verb* a shell casing into the die, but we prefer the verb "stroke."

Die *noun.* It's one of a set, made of steel, which forces the brass into a certain shape. After resizing, the expander die pushes the neck of your brass out so it is exactly the size of a bullet.

Shell holder This clips onto the top of the ram and accepts a shell casing in its slot.

FPS = Feet per second (velocity).

IMR = Improved Military Rifle, (Dupont rifle powder).

OCL or **COL** = Overall Case Length / Case Overall Length

PSI = Pounds Per Square Inch (pressure measure).

Let's consider some tools you need to put a rifle cartridge together. A powder measure drops the same amount of powder (supposedly) every time you flip the handle. A few things interfere with accurate powder dumps out of a measure, but you check any powder dump with a scale to weigh the powder accurately.

Rifle cases stretch because of the hot, high pressure squish on the brass outward when fired. The brass has to go somewhere, so the case grows---forward at the neck. A case length gauge makes sure a case isn't too long, and a trimmer cuts a little brass off the neck. After trimming, a hand reamer barbers the neck's leading edge. Of course the heart of the whole operation is a press, which holds and operates the dies. We like dies which require no case lubrication.

Reloading cartridges is like cooking. You first decide what food you'll make, then you get ingredients together. You then obtain tools and equipment and begin in a logical manner.

CHOOSING CASES

What about choice of cases? Many shooters specialize in OP's. That's the most popular case ever reloaded. OP stands for Other People's. The shooters who use them are easy to identify on the range; they are

Almost all manufacturers are afraid to put together a round for any military purpose. So you can't purchase bullets that go through anything. If you want a piercing bullet, you'll have to drill the tip of a full metal jacket and insert (press fit) a diamond needle from a phonograph. Probably, the diamond would penetrate what the bullet won't. Also, the military makes .30 caliber ammo in a variety of ways and color codes the bullet tips to tell you what a particular round will do. One of those colors is black, which means AP, or Armor Piercing. Pull the black tipped bullets and save them. In another reload, you can make holes in barricades and truly surprise those hiding on the other side. Tip: Weigh the bullet. From your reloading manual, choose a powder for that bullet weight. Need more velocity? Use Vihtavouri powder from Finland to cut down on chamber pressure. Then add more powder. Smile.

always giving you the vertical smile—-bent over, picking up everything they can find. You may be an OP shooter too, so separate your cases by brands and store them in lots. You don't need to buy a lot of expensive Tupperware for this. I use old tennis socks hung from nails with a little label. Winchester brass is popular; let's use that. As we prepare our brass, we will check for defects and discard accordingly. Here's the rule:

When in doubt, toss it out.

Measure all cases for minimum and maximum length, by running them through a Lee case-length gauge. You can do this quickly; set aside elongated cases for trimming with the Lee hand trimmer.

Let's cook up a rifle round—-a .308 Winchester because it's a military cartridge (7.62 mm. X 51) with lots of OP brass available. Also, when you look in the Hornady Manual, you'll notice a variety of loads similar to the 30-06. Though many cartridges equal or beat the .308 such as the .25-06, .270, 7 X 57mm., and the .280, none of them fit short actions.

> The flat part on the back of some bullets creates a vacuum so the bullet slows down fast. For long distances stay with a bullet shaped more nicely on the rear, such as a boat tail, or BT. Bullets sort of succeed like Hollywood starlets. It isn't the nose that counts so much, it's how well-reared they are.

Just like a cook, you need to ask some questions. How will you use this? Hunting? Defense? If defense, will you need penetration? Will you shoot long distance or short? What kind of animal will you hunt, which translates: How much should the bullet weigh and how much energy do you need to make the bullet expand?

CHOOSING THE BULLET

The weight of the bullet determines the powder charge, so choosing the projectile is what the cook does first. Long, heavy bullets go the distance, but you have to push them and the recoil can hurt.

A BULLET SHAPE NO-NO

Don't use pointed bullets in tubular magazines. With the points aimed right at the primer on the cartridge in front of it, a severe recoil can fire all six shots at once in the magazine. Find <u>the</u> bullet which is accurate in your rifle. Stick to it and stock up on them.

All modern jacketed hunting bullets are designed to give maximum penetration and expansion at their listed muzzle velocities. Most will expand at 1000 fps.

Let's suppose you hunt deer in a rough semi-bushy locale, with occasional 300 yard clear areas, which will be the maximum shooting distance. So we'll cook up 165 grain spitzers with enough powder to push it at 2,600 fps. for less muzzle blast & recoil. Note: Barrel and case life will be longer and you use less powder with a light load.

Find out what powder and primer to use so you get the best velocity and accuracy by reading your loading manual. You'll see a wide range of powders to use with each bullet weight. Where do you start? Look at the pressure level and the velocity yield for each bullet weight. For example, the maximum .308 pressure is 62,000 psi; that pressure pushes a 165 grain bullet at 2,670 fps. Hint: If you want long barrel and case life, avoid the high pressure loads.

> In 165 gr. weight, you can purchase BPBT's, Banana Peeler, Boat Tail. When the nose expands, the copper covering peels back like a banana and buzz saws into the target. Don's favorite bullet.

Keep records with color markers on the back of the brass and use the same marker on your reloading book page on which you record your cartridge load notes. Why? One special lot of cartridges you create may shoot better than a cruise missile—-or you may experience excessive chamber pressure.

If you observe any one of the following excessive pressure clues while shooting, stop and check things out.

Primer cratering or flowing back into firing pin hole.

Action hard to open.

Sticky or hard extraction of case from chamber.

Extractor marks visible on head of case.

Enlarged primer pocket.

Gas leaking around primer pocket.

Case head expansion.

Blown primer, pressure reaching 80,000 lbs.

Short case life.

WHAT COULD CAUSE THE TROUBLE?

1. Wrong data. Mark up your manual; you can always get a new one. Use a ruler and highlight the load you have chosen. Different colors might indicate the following.

Blue: Light load, cold. Green: Load tested and works well in your rifle; good to go. Amber: Caution, untested rounds in this lot, might fly wild. Red: Hot load; maximum pressure and velocity for this powder. Learn to read the data on one line in your manual. Draw a line with a ruler.

2. Wrong or mis-marked powder. This is a severe error. Talk about an elevator not going all the way to the top. . . The only guy I ever heard of who did this was a confused dyslectic, agnostic, insomniac. He kept staying up all night trying to figure out if there really was a doG.

3. Mixed powder. Left one kind in measure, added another. See #2!

4. Wrong bullet weight. Be really careful when you return bullets to a box. Place only one box down on the table at a time.

5. Wrong diameter bullet. Could you insert a .35 in a .30? Uh-oh.

6. Used a case which was too long. So your rifle bolt jammed the bullet up into the lands which made it act sort of like a barrel obstruction.

7. Bullet seating depth wrong; too deep into case compresses powder.

8. Powder scale set up inaccurately or not level.

9. Extra tight chamber and barrel.

10. Very hot weather. Arizona desert, cartridge in sun heat over 125°.

11. Very hot chamber and barrel (which will pre-heat powder and cause quicker burn).

12. Any change in reloading components. For example: Magnum primers raise chamber pressures. You only need magnum primers to ignite slow burning powders in large magnum cases to provide reliable ignition.

CHOOSING POWDER

The key is not how much pressure the weapon will take, but how much your brass will take. Your brass cartridge case is the weakest link in the pressure chain. Choosing a load always involves compromise. That's why you test more than one; you learn what powders and bullet weights are safest and most accurate in your rifle. What you want is good velocity at the lowest pressure possible. All manuals have "Starting" and "Maximum" loads listed. Begin loading by starting at the bottom and working up. That's the safest way to go.

Test, for example, IMR 4064 and IMR 4895 (on which standard primers work well). The .308 Winchester with 165 gr. bullet in front of 41 gr. of IMR 4064 will give you 2460 fps velocity at 42,900 psi. IMR 4895 powder starts out with a 40 gr. load yielding 2514 fps at 43,600 psi. Other powders create higher pressure and less velocity.

Once you choose a powder, load five cartridges each with different powder loads. Record the loading data on our forms. Everytime you change something, use a new page and key the page (identify it) so it matches the cartridge case. Sometime later test these loads at the range to look for accuracy in each five-shot group. Use our form to a keep a written record of results. We think it's highly important to tie loading data to performance. Also, check for signs of high pressure. Even if you're not on the edge of explosion, high pressures kill barrel life.

PREPARING THE CASE

Decap the brass by hand with a decapping punch. While the primer pocket is exposed, clean any residue out of the pocket. Now, clean the brass. We prefer tumblers. A one-hour treatment will make the case look new. Even more important, the primer pocket hole should be whistle clean. Any primer residue (called "cake") can keep the new primer from seating properly. With a power or hand

> "I've reloaded dirty brass often. It was often second class stuff. But it shot." Don

primer pocket cleaner, it only takes a turn to loosen primer residue. By holding each case in a clean rag while you clean out the primer pocket you also clean excess resizing lube from the case. If you use the reloading die to decap, clean the brass before you reprime the case.

SIZING

Either neck size only, or full-length resize the brass. On OP's, you have to full-length resize each case. Also, semi-automatics feed much more easily after a full-length resize because of the clearance space between the case and your weapon's chamber wall. Once a cartridge case has been fired in your rifle, it is fire-formed, which means it will fit snugly in **your rifle's chamber.** You can adjust your sizing die so it only resizes the neck, which means less stress on your brass. More important, the cartridge shoulder is up against the chamber shoulder and therefore holds the bullet nose exactly in the middle of the bore for accuracy. This is the best if you shoot long distances, such as 1,000 yards. Bullets not concentric with the bore can fly 2 MOA away from point of aim. At 1,000 yards, that's a 21 inch miss.

SETTING UP FOR FULL LENGTH RESIZE

Screw your resizing die into the reloading press head. Mount the .308 shell holder onto the press ram. After stroking (with hand on the press handle—-usually down) the ram up to its full height, screw the die down till it touches the face of the shell holder. Now back the die out a half turn. Put a little case lube around the case's neck. When you stroke the case up and then out of the die again, you can see how deep it went in by the lube mark on the case neck. The deeper you screw your die in, the further down the lube mark will be. Just before this mark touches the case's shoulder, stop. Stroking a case too deep into the die will crush the shoulder and ruin it. When you have it adjusted right, lock the die in place.

If you're resizing a straight walled rifle or pistol case run the die down till it touches the shell holder, then back it off 1/4 turn. When you resize a case, the ram should reach its highest position without much resistance. Don't force anything. Check out any abnormal resistance you feel as you stroke the press. Several things could go wrong, such as a collapsed case or a broken case rim.

Once you adjust the die, resize the cases. Put a small amount of lube* on your lube pad. Lightly roll each case on the pad just before you run it through the resizing die. Keep your cases separated as you proceed through each step. On the up-stroke you will be pulling a neck sizer out of the case and repriming the case. A little dry lube on the neck expander helps.

*I hate case lube. You dribble it on a stamp-like pad, roll your hands in it, then wind up touching primers with oily hands, which converts the primer to a dud. Duds not only don't fire, they make you lose confidence in your ammunition. You either have to pull them apart or toss them in the trash. The only practical use is to make all your friends in camp increase their blood circulation when you toss them into an open fire. This can happen also: You dimple your brass inward (hydraulic lock) from too much case lube, which by now has accumulated up inside your die. Finally, you leave goopy cases laying around and dust sticks to them. To avoid all this trouble: **Purchase carbide dies.** Don.

PRIMERS

Primers are explosive; handle with care. Keep them in original containers until ready to use. Petroleum products kill primers, so keep them dry and away from liquids. On progressive reloaders clean the primer feeder tubes often to avoid primer dust build-up and the explosions that sometimes follow.

Place a primer, striking surface down or anvil up, in the primer plunger. With case in shell holder, push the primer cup underneath the case and lift gently and firmly on the press handle to seat the primer. Make sure the primer is seated flush or below the case head. Result: Primed cases. You can store them or continue reloading.

For powder charging place all the primed cases in a loading block unless you have a progressive press. Items you need on the bench will be: Primed cases, one can of powder, one box of bullets all the same weight, loading blocks, powder scale, powder measure, seating die, four empty ammo boxes and press. That's all. Leave nothing, zero, junk laying around on your bench.

"More dynamite, Newt." Line from the movie, *Butch Cassidy and the Sundance Kid* spoken while paper money filled the sky. The line is appropriate when you consider the next topic.

HOW MUCH POWDER?

Set up the powder scale in a level area and use the adjustable leveling screw. Our first 5 cartridges will be loaded with 40 grains of IMR 4895 powder, so you set the balance weight at 40 and leave the tenth grain balance at zero. Attach your powder measure securely to a bench or other solid object and adjust it to throw 40 gr. of IMR 4895 consistently. Some measures have a dial to adjust the grain amount; others have a micrometer-like adjustment.

Try several powder throws by weighing them on your powder scale. Satisfied they are all consistent? Keep your powder reservoir full so you have the same weight in the reservoir. Throw all the charges the same way with equal force. Consistency is the key. Use your powder scale to recheck its charges no more than every 20th charge. Check again after you fill the last case.

Since a double powder charge can ruin your day, lets drop only one charge in each case. It's a good idea to count aloud as you run across the loading block. If your beneficiary needs money, drink alcohol before this procedure. Your life insurance may pay off double. After all cases are filled, inspect each of them with a pen light. Shine the light down in each powdered case to make sure you haven't dropped a double or perhaps bypassed one. Do that, and a primer only will lodge a bullet a little ways up the barrel. Fire another round behind it and you'll be amazed.

DROPPING POWDER FOR MAXIMUM LOADS

The natural thing for any American is to go for the max. In reloading, however, we don't recommend it. Max loads screw up brass and give barrels a short life. Still, we understand how some reloaders will just have to try. In addition, it's a good idea to know how in case you get into a situation where you need either max range or penetration.

You reach the max with any caliber in

> Don't seat a bullet too far down in the case neck and compress the powder. If you do that, pressures inside the chamber will build up like elephant gas.

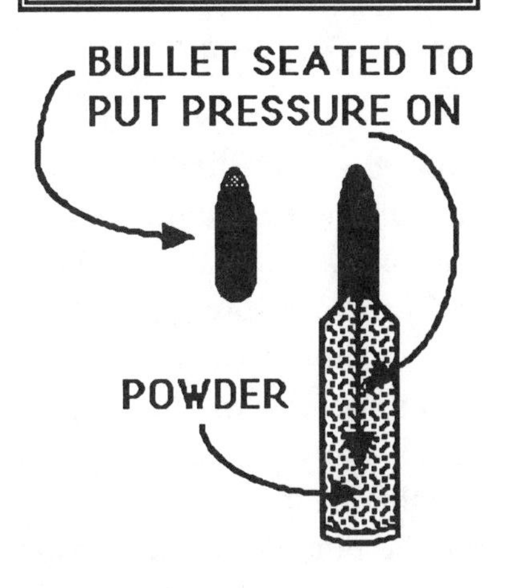

one of two ways: A. The case can't take any more powder. B. The brass can't take any more pressure. A is obvious. You'll inspect it with a pen light to check the powder level. To discover B, check for the pressure signs listed earlier. To avoid both problems, try Vihtavouri Oy powder. It will get you 30% more velocity with less pressure.

Don't trust your powder measure when loading to the max. Set up your measure to drop a light load, which you can drop into your powder scale pan. Hold the pan up tight against the measure, otherwise some powder will bounce out. Each charge should be a little light. Dribble powder into the scale with your trickler to get perfect weights. A powder funnel makes sure every granule goes into the case.

SEATING THE BULLET

The bullet seating depth is crucial for flawless weapon operation and accuracy. For best accuracy, a chambered cartridge's bullet should almost touch the weapon's rifling with the bullet nose exactly in the middle of the bore (concentric). Otherwise the bullet enters the bore slightly crooked—-and flies out the muzzle a bit erratically. Seat the bullet far enough into the case so the neck has a good grip on it—-and it has just a tiny bit of travel (.002" or .003" inch) before it slides into the rifling.

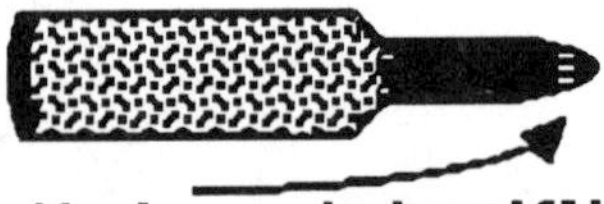

Marks made by rifling on blackened bullet tip

How do you adjust the seating die so it sets the bullet into the case just right for your particular rifle? You get close by placing a loaded round (commercial or store bought) in the shell holder, stroking the handle down to run the ram all the way to the top and gently screwing the seating die down until it touches the bullet. Now set a bullet of your own in a test cartridge with no primer or powder. Place the bullet on top of the cartridge case and guide it up into the seating die as you stroke the ram upward.

Mark the nose of the bullet with a dark highlighter or light a candle away from the loading area and blacken the tip of the bullet. Then chamber the round slowly and deliberately; try to feel for resistance. If you get no rifling marks on the bullet, back off the seating die so the bullet sticks out farther forward. Stop when chambering puts rifling marks on the bullet. Turn the seating die down just a little. Some rifles shoot better with bullets seated a few thousandths deeper; make five of those, too. Fire your different cartridge lots from a sandbag rest. When one seating depth groups well, use the same data from our loading form for future rounds.

SEAT FOR CONCENTRICITY

If you use a dial indicator on the side of the bullet as you turn it slowly, you'll notice any harmonic motion. Like a crooked pool cue rolling on a flat billiard table, the bullet will bounce up and down as you turn the cartridge. You may not have the equipment to do it yourself. NECO sells a kit that measures little wiggles in both case and bullet. Why should you check? Because some dies cause the bullet to be seated slightly crooked in the case. From the testing we have done, Lee dies are the most accurate you can get.

BE A BULLET POLISHER

For about a penny each, you can polish bullets with a NECO coat consisting of a chemical and carnauba wax. The next time you watch football, pay attention to a pass in which the ball wobbles and doesn't fly true. Bullets do that. But if you impregnate them they fly out of the muzzle with early bullet stabilization and land on target with significantly less bullet drop. As a matter of fact, treated bullets come out of the muzzle a little slower because they don't restrict powder burn. Adding more powder gets the velocity back up again. The big advantage is in bullet coefficient increase. Thus, the bullet speeds through the air and smacks.

After you seat each bullet, wipe the whole cartridge clean and make sure the bullet is tightly seated in the case. Put finished cartridges in an ammo box, keeping each different load in separate boxes. Mark each box with date and loading data. Keep track of the cartridges you load in a separate notebook. When you test your loads at the range, write down comments on its performance. After you've tried out several bullets, primers, powder, and cases, check your records and pick the most accurate load that produces the least chamber pressure.

Check performance against the nomograph you drew by shooting at various distances. You need to know that your mid-range trajectory height over the line of sight is correct. Satisfied with results? Stock up for a lifetime of shooting. If you are building ammo now for use much later, apply Gerorge and Roy's custom primer sealant to the primer cup with a toothpick. Use too much and you'll cover the flash hole.

On your finished boxes of ammo to be stored, tape one extra, randomly selected round to be fired at a later date. Just before you pack the boxes for a long trip, fire that one round as a test. I guarantee, you will love the sound of that one round. More than that, you'll be sure of the whole box.

Happy shooting.

IV Chapter 1

AIRGUNS

In a popular department store, recently, I watched two foreign speaking males talk about an airgun they intended to purchase. Although I'm a Gringo, I speak the language of these two gentlemen fluently. They confirmed what I already knew, these little silent shooters will destroy a car window easily and noiselessly.

Want to hear the worst endorsement in the world? More car radios and stereos have been stolen with an airgun than by many other methods. This is how they do it. They simply drive by, slow down, and pop the wind-wing with a pellet. Later, they drive by again, and if nobody has noticed the vandalism, they reach in, open the door, and close it gently behind them. In a matter of minutes, the wires are cut, the dash is destroyed, and the stereo is an insurance company report.

A five-stereo night is highly profitable. That accounted for how these two farm workers were dressed—-in $400 cowboy boots and another three hundred in clothes and hats. No wonder dance clubs for these people charge $20 per night. Living in America as car thieves is a great life.

Airguns have been around since the 1600's. Early types appeared much like their contemporary Flint Locks. When equipped with a rifled bore, they were very accurate and deadly. Lewis and Clark took one on their expedition across the North American Continent in the 1700's and it made quite an impression on the Indians. Many gun enthusiasts have classed them with the BB gun, OK for kids, but that's all. Our modern airguns are not toys and fill a gap in training, practice and small game hunting for several reasons.

PEST CONTROL & HUNTING

One person I interviewed bragged about how a neighbor of his would walk a dog, and the dog would fertilize his lawn. One day the dog apparently sat down on a bee or something, but it bit him just under the tail. That was the last of the fertilizer. Even with coaxing from the neighbor, the dog doesn't want to use that place as a bathroom anymore. For food, you can take out pigeons, squirrels, or small rabbits, as well as a variety of other edibles. Remember, no noise.

One airgun on the market called the "Arrow Gun" shoots crossbow arrows. This weapon can take large game, but I don't recommend it because of its high cost ($2000.00). Also, you need high pressure CO/2 bottles to operate it.

Shooting an airgun is ultra inexpensive. No matter what you pay for one, it won't take long to make up for what you might have spent on live ammunition practice. On the expensive airguns you will find the best trigger pulls, accuracy, workmanship and even scope mounting options.

TRAINING

Due to an airgun's silence, short range and low power, it's ideal for training new shooters or just keeping in practice. A shooting area can be set up anywhere, inside or out. All you need is a make shift, or commercial back stop, and 30 feet of range space. Officially, 33 feet is the range used in Olympic matches, however, shooting at a shorter distance can be just as rewarding. You just can't overlook the value of cheap practice on sight picture, trigger control and learning to trust your wobble. Airgun practice is much better than simple dry firing.

When we use an airgun for training, we call the gun the under-study. Under-study guns are the cheapo way of shooting to get you ready for the real McCoy. Thus, increase the weight of your airgun so it matches the weight of your hunting weapon. Also, add the weight so your airgun has the same balance.

The weight is easy. Simply subtract the airgun weight from the hunting rifle weight (loaded if you want to be precise) and add that amount to the airgun. Balance requires a little more finesse. Find the balance point of your rifle by resting it on a narrow board. Move the rifle back and forth until you locate the place where it stays put. Now, mark off that balance point in the same location on your air rifle. Divide the add-on weight factor by two and add weight to each side in places where you don't cause the balance point on the airgun to move.

Pests and small game are easy prey for the airgunner. Maximum effective range is 50 yards with a powerful airgun. If your hunting weapon is scoped, it's a good idea to scope the airgun also; use the same eye relief. You want to be fast in picking up targets through the scope. Also, you want to be able to measure your airgun wobble on our wobble grid and have it read the same as your hunting rifle would.

The power and accuracy of the airgun you choose will depend on how much money you spend. A good spring gun can cost between $200.00 and $1,000.00. You can expect velocities to run between 600 fps and 1000 fps, depending on how much money the weapon will cost. I recommend buying the most expensive field Model available. For one thing, pellets are dirt cheap. It won't take long to make up for what you would have spent on live ammunition practice. On the expensive models you will find the best trigger pulls, accuracy, workmanship and scope sight mounting options. Some airguns can last two lifetimes.

Airguns are inexpensive to shoot. A box of 500 high grade pellets costs between $6.00 and $8.00. That's a little over a penny each. You pay for no primer, now powder and you don't need cases.

Pellets are designed for plinking, target and hunting and come in calibers .177, .20, .22 and .25. You can store thousands of pellets in a very small area. To avoid oxidation, keep them in air tight containers and they will last forever. The majority of plinking and target pellets have flat or rounded points. Hunting pellets are either pointed or hollow pointed.

SPRING AIRGUNS

You cock spring airguns for each shot, either by braking the barrel down or with a side cocking lever. Cocking compresses a spring-loaded piston which compresses trapped air. This compressed air drives the pellet out the barrel.

Spring airguns have several advantages over pneumatic and CO/2 airguns. They give higher velocities, are more accurate, more efficient and quieter. They require less repair. They need no air storage cylinder or CO/2 bottle. They have no valves to wear out, need no pumping and are quicker to energize. They actually heat air on firing, drying out moisture in the barrel. On the other hand, pneumatic and CO/2 weapons produce rust-inviting condensation. Rather than barrels made of copper, brass or steel liners, the spring airguns are fitted with match grade rifled barrels for better accuracy.

DOWN WITH PNEUMATIC AIR

Americans make most pneumatic airguns. Single and multiple pump airguns and pre-charged bottle pneumatic air guns are available. Pre-charged bottle pneumatic guns are weak side as far as velocity goes. Pneumatic pump seals and valves wear out, which causes more maintenance than a spring gun.

CO/2 guns use carbon dioxide gas in small cylinders or large bottles. Their chief advantage over other airguns is repeat shots and ease of operation. Their disadvantages are many, but the main problem is supplying CO/2. Extended periods in the boon-docks will render a CO/2 gun worthless. Add to this their low velocity and power and you have an airgun unsuited for the survivalist.

LOOK INTO YOUR SHOOTING FUTURE

You can ride on air. A good spring-piston airgun is a valuable item in a survivalist's weapons battery. It will shoot better than you can hold. If you could mount one in a gun vise, you would find that all your shots at 30 feet would enter the same hole.

Get a high-grade field model and 100,000 pellets and you will be able to teach your grandchildren how to shoot with it. Airguns will last almost forever—-at least two lifetimes.

One word of caution, do not dry fire a spring airgun. Without a pellet the piston hits the receiver face and will damage the piston. Another word of caution: leave my car windows alone.

IV Chapter 2

BENCHES AND STORAGE

BENCHES

When you put a reloading bench together, make it stable. To do that, it's best to attach it to a wall, a tree or your motorhome. Of course you can buy all kinds of metal stands for work benches; these are rock solid when tied to a structure. Figure out how you want to reload. Sitting in a comfortable chair is one way to go, but many workbenches you may adapt for reloading are made for handymen who need to stand at the bench, move around, and muscle tools. The height of your bench dictates the reloading position in which you will spend hours. We recommend comfort. If you can do some operations away from the bench in a reclining chair, do that. Perhaps set up to use your powder scale from the standing position, and you sit down on a bar stool (bench height adjusted) for press operations.

The main part of the bench is the table top, and you need it strong and husky enough to hold presses firmly. Use Douglas Fir, 2 x 6, perhaps T&G (Tongue & Groove) screwed to three or four cross members.

RELOAD AT HOME?

You can build a permanent facility, designed and built by you with store-bought lumber. It ought to have a storage facility close by—-perhaps some used kitchen cabinets being replaced by a remodeling contractor. The usual place is in the garage, but attics are also workable. Watch the climate control. If you live in a hot area, partition off the attic and set an air conditioner to a thermostatic control. Air won't cost much in a confined area. Think about security. Your reloading area might go well in a home-made vault (plans in ***SECURE FROM CRIME*** by Huber). You don't want children or visitors playing with primers and powder.

IN THE WOODS

Another way to reload might be in the woods during vacation. In that event, you will either make a slab for a bench with a chain saw and place the slab on four hefty log stumps, or take a break-down bench with you. Cover the bench top with an awning, either umbrella-type or one made with canvass stretch over a wood frame. Fasten the bench to a tree so it doesn't wobble. If you leave reloading tools out overnight, cover them with a plastic garbage bag. Coat the moving parts of your presses (bare metal) with liberal doses of WD-40 or another rust preventive.

TOTAL PORTABILITY

In my forthcoming book, ***THE GREAT RV GETAWAY,*** I tell the reader a number of ways to earn a living while traveling. Since reloaded ammunition will probably be at a premium, think about being able to make it on the road. Most motor homes have a receptacle trailer hitch. Make a bench platform a slide in bar to fit. The same would work behind any other tow car or truck. Wing nuts will bolt your press to the bench platform. Get a tan while you create some of your best.

STORAGE AND CORROSION

Many reloaders sign off with the letters: KYPD, which means: Keep Your Powder Dry. To write all the appropriate letters, it should read, KYP-DCVH, which means Dry, Cool, (in a) Vacuum and Hidden. Four components make up the modern metallic cartridge. The metallic case, primer, powder and projectile. Keep all four items dry and in a cool environment. It's a good idea to read the regulations written by local, state and federal bureaucrats. Once in a while they come up with a safe idea.

Both new and reloaded ammo can corrode. The combination of contaminated brass, moisture accumulation, the powder and air can cause oxidation and electrolysis to both hand-loaded and commercial ammo. Brass corrodes. Weather conditions speed up oxidation and electrolysis to shorten case life if not stored or cleaned properly. Wash your cartridge cases with a commercial case cleaner, or just plain soap and water, before storing them. You need to remove any residue from powder or primer. Again, store cases as you would loaded ammo— in air tight ammo cans. A large G.I. mortar ammo can hold 17, one-pound powder cans. The plastic 25mm.

ammo box will hold 6 to 10, one-pound powder cans, and the .50 caliber ammo can will hold 6 to 8, one-pound powder cans.

Smokeless powder cabinets should be made of insulated materials and with weak walls, seams or joints to provide an easy means of venting. Powder should be stored in unconfined small quantities. Have more than one powder locker and keep them separated. Storage life for Ball powder is 50 years, and 20 years for extruded and flake powders. A cool storage place is a must for powder; excessive heat will destroy it.

Primers are igniters, much the same as detonators. Buy them in 1000 and 10,000 lots, and don't store them all together. Keep them away from your powder, because primers are very explosive.

Normally, you set primers off with a blow, but friction or electricity can do it as well. So handle with care and keep them in original packing and in a cool place to avoid deterioration.

Metal military ammo cans provide the best ammo storage. They're air and water tight, easy to handle and stack. Depending on the ammo size, you can store up to 1000 rounds in a .50 caliber can. Loaded ammo is considered safe and is not an explosive. Mark the cans; you can use a highlighter or a dyno-labeler. Refer to the bullet lots for which you keep records in your rifle book.

If you really want to hide it, dip your can in tar and bury it. If you're lucky enough to obtain a sealed tin of military ammo, resist the temptation to open it. Keep it sealed and it will last as long as you live.

HOW MUCH AMMO IS ENOUGH?

How much ammo should you store? No limit. If you load only for self, multiply your annual ammunition allotment by the number of years you'll be breathing and get busy. Think about sales, however. Ammo may become a precious commodity someday. For your own needs—-you'll shoot more shotgun ammo than rifle ammo. Add to this any trap or skeet shooting and your annual ammo requirement starts climbing. Pistol shooters burn up a lot of ammo because it takes practice to stay proficient.

Cartridge cases are made from plastic, brass, ~~paper, steel,~~ and ~~aluminum.~~ Don't use these last three. Plastic shotgun cases are durable. The brass case is by far the most widely used for pistol and rifle cartridges.

If you store black powder, be a lot more careful. Perhaps a bullet proof, 5/8" steel-sided box, lined with 3/8" plywood, which you keep locked is a good idea. This is just one reason why black powder weapons are second rate for the survivalist.

BULLET STORAGE

Bullets are made from copper, lead, or steel. All three can be contaminated by climatic conditions. Oxidation and rust can attack, so keep them in air tight containers. Store bullets in original containers, all marked with type, size and weight. Label your cast bullets clearly with a pen so you can read the labels a few years down the road. Since labels deteriorate, always check your bullets for size and weight before using. Never mix up loose bullets.

CARTRIDGE STORAGE

You may build several 5 cartridge sets during the winter, and not get a chance to test them until the weather turns. Remember, you color coded the rounds the same as the loading form we supply in this book. Once you have a super load which performs really well for you and your gun, however, you will be building boxes of the stuff. Simply use the form we provided for your rifle book, and shrink it so it fits a cartridge box. Then glue the labels either the plastic ammo containers you own, or the cardboard containers the brass or ammo comes in.

MIX 'N MATCH

Choose mix 'n match loads to cover a variety of situations from the boxes you marked. Shotshells can be extracted, one from each box, to make up a complete set. Perhaps code rifle bullets for short and long range by marking the tips. Black, of course, is for armor piercing. Simply load five different rounds in an ammo wallet and you go into the field prepared for just about anything.

No matter where you are or how you travel, putting the components of a fine cartridge together will be easy. With decent storage on the road or at home, you will always have enough components to keep a ready supply of super ammunition.

APPENDIX I
MAIL ORDER GUNSMITH SUPPLIES

As you may know, Don's books have been selling in unshops and Army Navy Stores for over a decade. Recently, on heard Republican Senator Orin Hatch (Utah) talk about 'linton's Omnibus Crime Bill. That bill banned 19 assault rifles. e said that off that bill, the ATF had built a list that banned more ıan 175 different firearms.

Knowing how Democrats think about taxing, then ɔending to create a new class of poor (welfare) voters, and nowing that they can't have all the power they crave with guns vailable, you can expect to see a clamp down on ammuntion. o, we think this list is important enough to print in both volumes f this series. Here's a few places you can obtain mail order nmo and components. Keep off the mailing lists if you even ıspect the ATF might confiscate your name.

Brownell's 200 South Front St. Montezuma, Iowa 50171 Phone 515-623-5401 Fax 515-623-3896 Gunsmithing supplies, tools, books, and gun parts.

Gander Mountain P.O.Box 248, Wilmot, WI. 53192-0248 1-80(558-9410 has a full line of reloading equipment and components, pl loaded ammo, shooting supplies, and sights.

Huntington RCBS 605 Oro Dam Blvd. Oroville, CA 94965 1-80(533-5000 I've used RCBS dies for years. Huntington RCBS has full line of reloading equipment and special order reloading dies, ar reloading components.

Midway 5875-DW. Van Horn Tavern, Columbia, MO 65203 800-243-3222 has a full line of reloading supplies and equipmen They also holsters, sights, and rifle stocks.

NECO, 1316 67th St. Emeryville, Ca 94608 510-450-0420. Fire lapping (patented process) for rifle barrels and coating systems to gai in bullet coefficiency.

Sinclair International 2330 Wayne Haven St. Fort Wayne, Indiar 46803 Phone 219-493-1858 Fax 219-493-2530 has a special line (special reloading equipment, as well as other brands.

Old Western Scrounger 12924 HWY A-12 Montague, CA 9606 Eley, Ca Fax 916-459-3944 1-800-877-2666 Service 916-45 5445 has a full line of reloading equipment and supplies, for rifl pistol, shotgun, and cannon. He has ammo from Fiocchi, Gec Weatherby, Rottweil, as well as domestic and hard to find items.

RELOADING SUPPLIERS

ACTIV Ind,1000 Zigor Rd. P.O.Box 339, Kearneysville, WV 25430
American Products Co., 14729 Spring Valley Rd. Morrison, IL 6127(
B-Square Engineering 2708 St. Louis Ave Ft. Worth TX 76109
Ballistic Products, 2105 Shaughnessy Circle Long Lake MN 55356
Barnes Bullets, Inc., P.O.Box 215, American Fork, UT 84003
Beeman Airguns 5454 Argosy Dr. Huntington Beach,CA 92649
Belding and Mull, P.O.Box 428,100 N.4th St.Phillipsburg, PA 16866
Benedict, Drew Mr., P.O.Box 368, Oroville, CA 95965
Birchwood Casey, 7900 Fuller Rd., Eden Prairie, MN 55344

Bitterroot Bullet Co., P.O.Box 412, Lewiston, ID 83501
Brownell's Inc., RR 2, P.O.Box 1, Montezuma, IA 50171
C & D, Cherokee Ind Pk, 309 Sequoya Dr. HopkinsVille, KY 42240
C-H Tool and Die Corp., 106 N. Harding St., Owen, WI 54461
Dewwy, J. MFG. Co., P.O.Box 2014, Southbury, CT 06488
Dillon Precision Prod 7442 E. Butherus Dr. Scottsdale, AZ 85260
4-D Die Co., 711 N. Sandusky St., Mount Vernon, OH 43050
Forster/Bonanza Products, 82 E. Lanark Ave, Lanark, IL 61046
Freedom Arms, P.O.Box 110, Freedom WY 83120
Fremont Tool Works, 1214 Prairie St. Ford, KS 67842
Camco Gunsmithing, 3509 Carlson Blvd., El Cerrito, CA 94530
Hercules Inc. Hercules Plaza, Wilmington, DE 19894-0001
Hodgdon, P.O.Box 2932, Shawnee Mission, KA 66201
Horizontal Archery, R2,37432-5th Ave., Sardis, Ohio 43946-9601
Hornady Mfg. Co., P.O.Box 1848, Grand Island, NE 68802-1848
Huntington Die Specialties, 601 Oro Dam Blvd. Oroville CA 95965
Jones,Custom Prods RR 1, PO Box 483-A Saegertown, PA 16433
K & M Services, P.O.Box 363, 2525 Primrose Ln., York, PA 17404
Lage Uniwad, Inc., P.O.Box 446, Victor, IA 52347
Lee Precision, Inc., 4275 HWY."U", Hartford, WI 53027
Ljutic Industries Inc., 732 N.16th Ave., Yakima, WA 98902-509
Lortone, Inc., 2856 N.W. Market St., Seattle, WA 98107
Lyman Products, Rt. 147 West St., Middlefield, CT 065455
Marquart Precision Co., P.O.Box 1740, Prescott, AZ 86302
Neco, 1316-67th St., Emeryville, CA 94608
Old West Gun Room, Inc., 3509 Carlson Blvd., El Cerrito, CA 94530
PolyWad, Inc., P.O.Box 7916, Macon, GA 31209
Ponsness-Warren, P.O.Box 8, Rathdrum, ID 83858
Quintics Corp., P.O.Box 29007, San Antonio, TX 78229
RCBS, Blunt, Inc., 2299 Snake River Ave. Lewiston, ID 83501
R.D.P. Tool 49162 McCoy Ave., East Liverpool, OH 43920
Redding, Inc., and Saeco Prod., 1089 Starr Rd., Cortland, NY 13045
Reloading Specialties, Inc., P.O.Box 1130, Pine Island, MN 55963
Remington Arms Co., 1007 Market St.,Wilmington, DE 19898
Rock Crusher Press,5324 Manila Ave., Oakland, CA 94618

Rooster Laboratories, P.O.Box 412514, Kansas City, MO 64141
RWS Ammo, 12924 HWY A-12, Montague, CA 96064
Sinclair Int., Inc., 718 Broadway, New Haven IN 46774
Star Machine Works, 418-10th Ave., San Diego, CA 92101
Sturm Ruger Co., Inc., Lacey Place, Southport, CT 06490
Tru-Square Metal Prods.,Inc., P.O.Box 585, Auburn, WA 98002
U-Load, Inc., 14952 Martin Dr., Eden Prairie, MN 55344
Vitt/Boos Aerodynamic Slug, 2178 Nichols Ave. Stratford, CT 06497
Weatherby, Inc., 2781 Firestone Blvd., South Gate, CA 90280
WG Gunsmiths 5324 Manila Ave., Oakland, CA 94619
Wilson, L.E. P.O.Box 324, 404 Pioneer Ave. Cashmere, WA 98815
Winchester Group, Olin, 427 N. Shamrock St., East Alton, IL 62024

GLOSSARY

ACCURIZE To make accurate. This is what shooters commonly do to rifles and handguns to make them shoot right on target.

AP Armor Piercing. A black-tipped bullet used by the military.

ASAT All Season, All Terrain. Commercial camouflage cloth not only colored correctly, but lighweight to blow in the breeze.

BOOM The sound a weapon makes when it fires. Critical for discovering the distance away a sniper is located.

BUTTPLATE The part of a gunstock which fits into your shoulder.

CHARGE Also powder charge. Goes by weight in grains only.

COMPOSITE STOCK. Not wood, but hi-tech resin. Warps in heat.

CONCENTRIC Describes circles with same center. Bullet-to-bore concentricity is important. Without it, bullets enter rifling off center.

CRACK. The sound a bullet makes as it passes nearby.

CRISP. Sudden. Describes a trigger that snaps without warning.

DOUBLE AUGHT. (say, ought) 00. Aught is an old term for zero, and double aught buck is 00-buckshot, each .33 inch in diameter.

DRY FIRE. To snap the trigger on an unloaded weapon. It's an excellent way of developing an educated trigger finger.

EJECTA. A general name covering all of the lead, sabots, slugs and other stuff you shoot out of a shotgun, including salt crystals, (stings)

EXTRACTOR On the bolt face, the springy lip which grabs the spent shell casing and pulls it out of the chamber.

FIRE-LAP To fire a special bullet with polishing grit, and lap the bore.

FLECHETTE A kind of shotgun shell you put together with # 10 wire, (large paper clip) so that the shot charge will only penetrate humans and not walls at close range. Safety first. . .

FOREARM. Front part of the weapon stock under the barrel.

HOLEY With new holes, made by bullets.

INCOMING Bullets others shoot at you.

INTEL. Short for intelligence, it refers to all you need to know about a military or tactical situation.

LANDS AND GROOVES The spiral inside of any rifle or handgun has grooves cut into the barrel. The lands remain to grab the bullet and cause it to spin as it leaves the barrel.

LONG GUNS Shoulder fired weapons; not handguns.

MATCH GRADE Ammunition made with closer tolerances. Sure.

MOA's Minutes of Angle. There are 60 of these in one degree, and this is the common way we measure small arms ballistics movement, both up and down & left and right.

MONTE CARLO A build up on (mostly) rifle stocks which causes the

shooter's eye to line up perfectly with sights. On a shotgun, we would like to see a Monte Carlo cheekpiece filled with gel.
NOMOGRAPHIC Describes rise and fall of bullet in flight.
OP Other People's, as in brass. What someone else shot and you pick up to take home and un-shoot, or reload. Don's favorite.
RANGE The distance from the shooter to a target.
RETICULE Pronounced, "retikle." The device inside a scope which enables you to aim, such as cross hairs, dot or post.
RUBBING COMPOUND Brownells offers a compound paste with grit in it. In between two rough surfaces, it makes both smooth.
SABOT A bullet smaller than your barrel size wrapped up in plastic which separates as soon as the bullet clears the muzzle.
SAYONARA Japanese word for good-bye, frequently used by rifle shooters in long range combat against lesser weapons.
SCATTERGUN Slang for shotgun, because it scatters shot.
SPANDOFLAGE A stretch net of camouflage material one wears over the face.
SEMI-AUTO Feeds without having to put a new round in the chamber, but you have to pull the trigger each time you want to make it fire.
SOFT POINT On a jacketed bullet, the exposed lead tip of soft material which helps the bullet to mushroom inside the target body.
WHOMP Bullet energy at target. Same as smack, knock over, etc.
WHUMP What a rifle barrel does when a bullet twists out of it at high speed. Sort of like twisting two ends of rubber hose in opposite directions.
WIND VALUE Not just the wind alone, but how strong it is on your particular bullet. Measured in Minutes of Angle, or MOA.
ZERO *verb and noun.* To zero your sights, you adjust them so the strike of the bullet coincides with your point of aim. You change your zero when you re-adjust, or you lose your zero when you drop your rifle on rocks so it lands scope first. Ouch.

.308, .22, .250, etc. Calibers measured in hundreths of an inch in diameter across the muzzle of a weapon.

INDEX

When in doubt toss it out. New brass cost a dime, new weapons cost hundreds, new eyes cost millions.

K.Y.P.D. - Keep Your Powder Dry and cool also. You can store powder that way for 20 years.

Author David B. Smith with mickey mouse ears is a quarter-shooter with this .25 caliber wildcat bullet launcher he designed and built. He designed special chamber for this necked-down super blaster.

Path Finder Publications

1296 E. Gibson Rd, Suite E-301, Woodland, CA 95776

Check Selections for other books we publish

Tear out or copy this page and send with order

AMMO FOREVER II The Complete What To Shoot & How Manual, Book II, Handguns. **$14.95** Any handgun you buy is only worth half until you improve it. After that, learn to make super ammo and shoot it better than the Lone Ranger. Includes combat, tactics cover and shoot-out techniques. Available Feb '95

Everybody's Knife Bible. **$12.95.** Landmark book on new, outdoor knife uses. Outdoor Life Book Club selection. *American Survival Magazine* said, "...16 of the most innovative and informative chapters on knives and knife uses ever written." Just under 30,000 now in print.

Never Get Lost. (Also known as: The Green Beret's Compass Course) $9.95. Best and most simple land navigation system anywhere. We've sold over 28,000. **Throw out your maps**; go anywhere you want, then bee-line back to your starting point without having to back track. System also works equally well in darkness. You can send two separate vehicles in two separate directions with no maps and have them meet anytime, anyplace.

24 + Ways to use your Hammock $4.95 From crabtrap, to gun rest, to camouflage ghillie suit. No kidding; we really figured them out.

Everybody's Outdoor Survival Guide. **$12.95.** More innovations. Teaches exclusive outdoor know-how found nowhere else. Long range and defensive platform accuracy shooting. Animals for survival. Hand to hand combat principles, water purification, plus a lot more.

Great Livin' in Grubby Times. **$12.95** More advanced firearms and survival, including weapons selection and team defensive shooting. Our big seller, now in 3rd edition. Contains popular info from Green Beret Brian Adams on Escape and Evasion.

SECURE FROM CRIME, How To Be Your Own Bodyguard **$14.95** Tips and tricks never becoming a victim. Over 15,000 sold since April '93. The book you can't live without, because without this book, you might not live. Contains new, advanced systems to protect yourself from the rage of America.

Use other side as cash and subtract $3---or **$4 with U.S. Postal money order.**

Add $1.35 for Shipping & Handling.

DISCOUNT COUPON

Tear out this page or copy it on a Xerox. Use this discount coupon to order any of our books on the other side of this page.

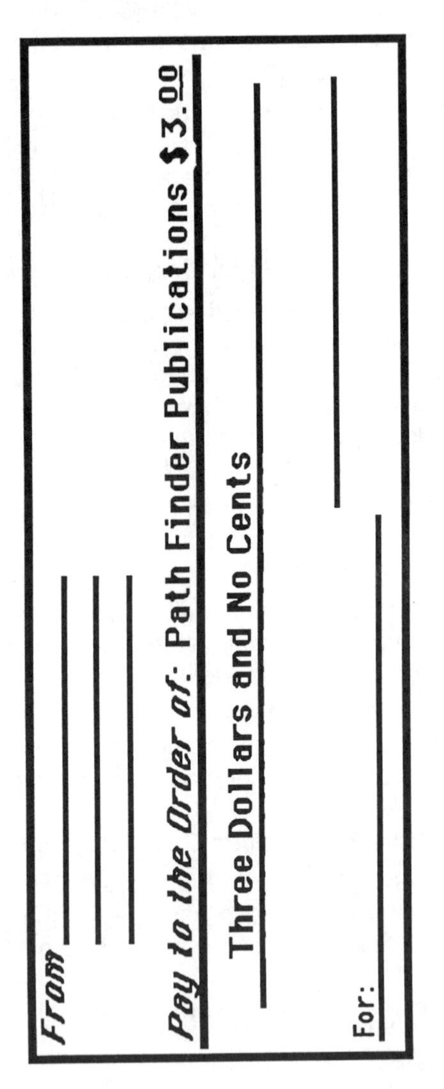

Path Finder Books are guaranteed. If your book suffers water damage, fire, or consumption by goat, we'll send a new one for half price.

When ordering, add $1.35 for shipping and handling after deducting the discount amount on the coupon to the left.

Dealers and book stores: Please accept this coupon on any of Path Finder's books. We guarantee to redeem this for you in keystone product.

Path Finder Publications
1296 E. Gibson Rd, E-301
Woodland, Ca. 95776

THE GREEN BERET'S COMPASS COURSE

The New System For Staying Found In The Wilderness

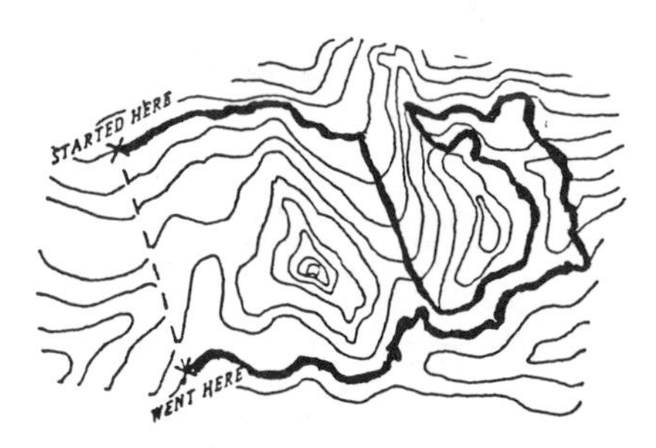

Suppose you walked the path of this black line on this contour. Now choose: Would you like to back-track home or take the straight-in short cut?

MAKE YOUR OWN COMPASS

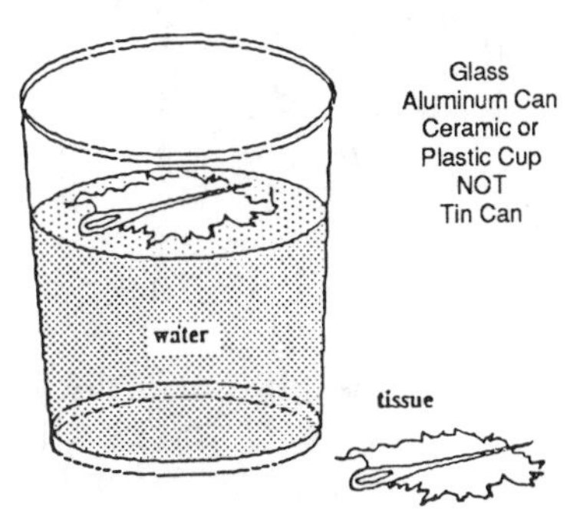

Find out how much time is left till sundown using the fat finger trick.

Tissue will float on water until it soaks. Magnitized needly will turn and point to magnetic north.

Fast becoming the land navigation system-of-the-future for all outdoorsmen, this book keeps you from ever getting lost in the woods. The unique PAUL method requires no maps, and shoots you straight back home by the shortest route possible--a straight line.

By Pathfinder

New-Method Books For Real Outdoorsmen